Global Warming and Social Innovation
The Challenge of a Climate-Neutral Society

Global Warming and Social Innovation
The Challenge of a Climate-Neutral Society

Edited by
M.T.J. Kok, W.J.V. Vermeulen, A.P.C. Faaij, D. de Jager

Earthscan Publications Ltd
London • Sterling, VA

First published in the UK and USA in 2002
by Earthscan Publications Ltd

ISBN: 1 85383 945 0 paperback
         1 85383 944 2 hardback

Typesetting by Marjan Kramer
Printed and bound in the UK by Creative Print and Design Wales, Ebbw Vale
Cover design by Danny Gillespie

For a full list of publications please contact:

Earthscan Publications Ltd
120 Pentonville Road, London, N1 9JN, UK
tel: +44 (0)20 7278 0433
fax: +44 (0)20 7278 1142
email: earthinfo@earthscan.co.uk
http://www.earthscan.co.uk

22883 Quicksilver Drive, Sterling, VA 20166-2012, USA

Earthscan is an editorially independent subsidiary of Kogan Page Ltd and publishes in association with WWF-UK and the International Institute for Environment and Development

A catalogue record for this book is available from the British Library

Library of Congress Cataloging-in-Publication Data

Global warming and social innovation : the challenge of a
climate-neutral society / edited by Marcel Kok ... [et al.].
    p. cm.
Includes bibliographical references and index.
    ISBN 1-85383-944-2 (hardback) -- ISBN 1-85383-945-0 (pbk.)
1. Global warming--Social aspects. I. Kok, Marcel, 1968-
QC981.8.G56 G58147 2002
304.2'5--dc21
                        2002013821

This book results from a collaboration initiated by the Dutch National Research
Programme on Global Air Pollution and Climate Change.
P.O. Box 1, 3720 BA Bilthoven, The Netherlands
    tel: +31 (0)30 274 29 70
    fax: +31 (0)30 274 44 40

This book is printed on elemental chlorine-free paper

Contents

Foreword

This book is the result of a collaborative effort by researchers in The Netherlands within the context of the Dutch National Research Programme on Global Air Pollution and Climate Change (NRP). It starts from the assumption that the problem of climate change will require long-term solutions leading to deep reductions in emissions. The authors have tried to develop ideas on how a 'climate-neutral society' might be realized. We have attempted to go beyond the traditional boundaries of academic disciplines by using a back-casting approach, and we hope that the final outcome of this process will contribute to the ongoing debate about new strategies to combat the problem of climate change.

As editors, we would first like to thank all the authors for their contributions to this book and their willingness and enthusiasm to discuss and review each other's work. We would also like to thank Professor L.J. Nilsson (Environmental and Energy Systems Studies, Department of Technology and Society, Lund University, Sweden), Professor R. Socolow (Princeton University, Centre for Energy and Environmental Studies, USA) and A. Scott (University of Sussex, UK) for their constructive and in-depth reviews of the book. Their contributions were very valuable in improving both the individual chapters and the book as a whole.

Last but not least, we wish to thank Ottelien van Steenis (NRP) for her organizational support in the entire process of making this book what it is, Hugh Quigley (Effective English) for his skilful editing and Marjan Kramer (RIVM) for the graphic design.

List of figures, tables and boxes

Boxes

List of acronyms and abbreviations

CHP	Combined Heat and Power
EU	European Union
FCV	Fuel Cell Vehicles
GNP	Gross National Product
ICT	Information and Communication Technology
IPCC	Intergovernmental Panel on Climate Change
MIPS	Material Intensity Per Service Unit
NEPP	National Environmental Policy Plan
NGO	Non Governmental Organization
NRP	Dutch National Research Programme on Global Air Pollution and Climate Change
OECD	Organisation for Economic Co-operation and Development
PPMV	Parts Per Million Volume
PV	Photovoltaic
R&D	Research and Development
RIVM	National Institute for Public Health and Environment
SRES	Special Report on Emission Scenarios (of IPCC)
UNDP	United Nations Development Programme
UNFCCC	United Nations Framework Convention on Climate Change
WTO	World Trade Organization

Chapter 1

Towards a climate-neutral society

Marcel Kok, Walter Vermeulen, André Faaij and David de Jager

Introduction

Human-induced climate change presents society with a long-term problem. The challenge is to find ways of reducing CO_2 emissions in the short term that also enable society to reach long-term goals. Over the last 20 years, it has become increasingly clear that human activities are adding to natural changes in the climate. The enhanced greenhouse effect caused by emissions of greenhouse gases related to human activities is a major factor in currently observed climate changes, as well as in climate change that is expected to occur in the coming centuries (IPCC, 2001a).

The long-term character of climate change (having an effect for up to several centuries) is one of the key features of the climate change problem. However, dealing with long-term problems is generally difficult for modern societies, in which political agendas and practices tend to focus on the next shareholders' meeting or election rather than on the needs of future generations. If we are to minimize the human contribution to climate change, society will have to move towards towards a 'future of low greenhouse gas emissions' or in a 'climate-neutral direction'.

If society wants to realize what we refer to in this book as a 'climate-neutral society' (i.e., a society that minimizes its negative impact on the climate system and its contribution to climate change), substantial reductions of up to 60–80 per cent (compared to the 1990 levels) in the greenhouse gas emissions of industrialized countries will have to be achieved before the end of the century. For Western societies, this implies a major shift in current and expected trends in emissions – emissions that are expected to rise in the absence of stringent sustainability and/or climate policies (IPCC, 2000; UNDP, 2000). Achieving such drastic reductions in emissions will be no mean task since societies – particularly Western societies – rely heavily on the availability of relatively cheap fossil fuels and the idea that the atmosphere is a free common good into which greenhouse gases can be emitted. As the Intergovernmental Panel on Climate Change (IPCC) states, 'the successful implementation of greenhouse gas mitigation options needs to overcome many technical, economic, political, cultural, social, behavioural and/or institutional barriers which prevent the full exploitation of the opportunities of these mitigation options' (IPCC, 2001c). 'In the industrialized countries,' continues the IPCC, 'future opportunities lie primarily in removing social and behavioural barriers' to these technical solutions. But what if merely removing barriers is not enough for realizing these drastic reductions? Perhaps societal 'trend breaks' (substantial shifts in societal developments) are also necessary if we are to achieve a climate-neutral society.

The overall objective of this book is to explore *the need for societal trend breaks to realize long-term substantive reductions in the emissions of greenhouse gases and to assess the prospects for realizing such trend breaks in society.*

This book analyses both the long-term prospects for and short-term implications of the transition towards a climate-neutral society from different perspectives. In so doing, climate change will be placed within the wider context of sustainable development. The basic premise is that given the current situation in climate-change policy and the direction we may have to take in order to realize a climate-neutral society, it will probably also be necessary to change several social trends in a climate-friendly direction – as a pre-condition for decreasing emissions.

In this book, The Netherlands serves as a case study for questions and problems of a similar nature that will arise in other industrialized countries as they are confronted with the challenge of developing long-term domestic climate policies. The main focus of the book is the reduction of greenhouse gas emissions and the possibilities for domestic action within The Netherlands. The impact of climate change on humans and nature is outside the scope of this book, as are possible adaptation strategies.

Some chapters discuss the Dutch situation; other chapters reflect the debate in The Netherlands on strategies that in principle are also relevant for other countries. It is clear that besides domestic action, the so-called flexible or Kyoto Mechanisms (joint implementation, clean development mechanism and emission trading) will play an important role in future reduction strategies. With these instruments, international flexibility is introduced within climate policies, to achieve cost-effective emission reductions. The portfolio of climate policies will always consist of both domestic options and the use of the Kyoto Mechanisms. But given the focus on domestic strategies within The Netherlands, the discussion of the Kyoto Mechanisms will be limited in this book.

Much of the research presented in this book was carried out as part of the Dutch National Research Programme on Global Air Pollution and Climate Change, a strategic research programme in The Netherlands that aims to contribute to the development of long-term climate-change policies. This book is an effort to bring together some of the results of this programme in a coherent framework.

The necessity of a climate-neutral society: long-term challenges

Internationally, there is a growing acknowledgement of the need to achieve emission reductions beyond what is agreed upon in the Kyoto Protocol and of the challenges involved in doing so. This awareness arises from a growing body of scientific evidence, which shows the human influence on the climate. The most recent IPCC report (IPCC, 2001a), for instance, states that there is new and stronger evidence that most of the warming observed over the last 50 years is attributable to human activities. The result of this anthropogenic climate change is that society is now forced to reduce its emissions of greenhouse gases into the atmosphere. The United Nations Framework Convention on Climate Change (UN FCCC) states that the world has *'to stabilize the greenhouse gas concentrations in the atmosphere at a level*

*that would prevent dangerous anthropogenic interference with the climate system. Such
a level should be achieved within a time frame sufficient to allow ecosystems to adapt naturally to climate change, to ensure that food production is not threatened, and to enable
economic development to proceed in a sustainable manner'* (UN FCCC, art. 2).

What is regarded as a desirable level is ultimately the outcome of a societal
and political decision-making process. In Figure 1.1, reasons for concern about
future temperature changes are given, based on research on the impact of climate
change (IPCC, 2001b).

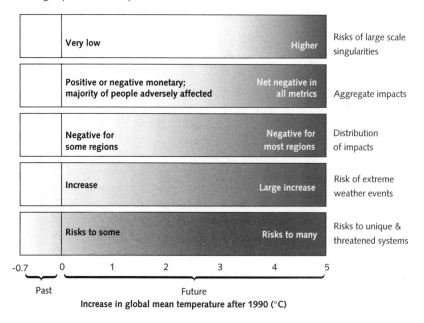

Figure 1.1 *Reasons for concern about the expected consequences of climate change (IPCC, 2001b)*

Ultimately, to achieve a climate-neutral society, the concentrations of greenhouse
gases in the atmosphere will have to be stabilized at a new equilibrium, where natural uptake by the oceans and biosphere equals global emissions. This implies that
global emissions will have to be cut by half (compared to 1990 emission levels).
Historically, industrialized countries have caused most of the observed increases
in greenhouse gas concentrations in the atmosphere. Taking equity considerations
into account, one could argue that the industrialized world has to make the greatest reductions in future emissions in order to leave 'space' (in greenhouse terms)
for development in the South.

Based on several points of departure, such as the maximum allowable temperature change per decade, The Netherlands and the European Union have stated that the concentrations of greenhouse gases should be stabilized before the end
of the 21st century at a level well below twice the pre-industrial level. The number
of 450 parts per million of volume (ppmv)[1] is often used to quantify a desirable

---

[1] The pre-industrial level of CO_2 concentrations was 280 ppmv and the level in 2000 was 368 ppmv.

level of CO_2 emissions. The absolute reduction in emissions necessary to realize this level (or any other level) depends on how the world develops. IPCC (2000) devised four future worlds or storylines and, subsequently, six emission scenarios, which reflect different developments in population, economy, energy, technology and land use. These so-called *SRES* scenarios are a combination of global versus regional orientations and an economic versus an environmental orientation. Figure 1.2 shows the different baselines (solid line) and emission-reduction pro-files (dotted line) of the world greenhouse gas emissions up to the year 2100 that realize a stabilization level of 450 ppmv CO_2 in the four future worlds of the IPCC SRES scenarios.

The long-term climate challenge for industrialized countries is to transform their carbon-intensive industrial society into a climate-neutral society, and at the same time work with the South in implementing sustainable development there while avoiding large-scale emissions during their economic 'take-off'. The transi-tion towards a climate-neutral society may seem a daunting task, but it is not beyond the bounds of possibility. Chapter 2 of this book presents two visions of how an 80 per cent reduction in emissions (compared to 1990 levels) can be achieved by 2050 in The Netherlands' energy system. The main building blocks of the solutions applied in these two visions are improvements in efficiency, the introduction of sustainable energy sources, reduction of energy demands and CO_2 storage. The two visions are based on the future worlds in the IPCC SRES scenar-ios A2 and B1 and employ different sets of measures and technologies. They are

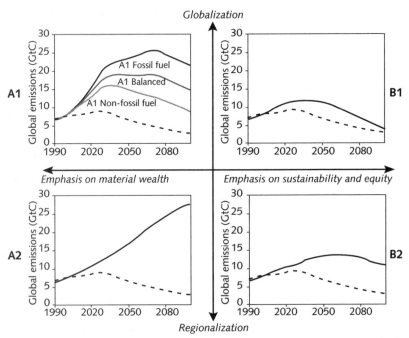

Figure 1.2 *Emission profiles (dotted lines) needed to achieve stabilization at 450 ppmv for the different IPCC scenarios (solid lines). Based on analysis with the RIVM FAIR model (Berk et al., 2001)*

meant to sketch a spectrum of directions that might lead to the development of a climate-neutral society. In this respect, Chapter 2 provides the reader with a starting point for considering possible technological routes of dealing with the long-term challenges of domestic climate-change policies.

Current climate-change policies

In the context of this book, the current policies on climate change and the short-term dimensions of long-term routes to emission reductions are also important. The current situation is basically the ground on which innovations have to flourish and emission trends have to be reversed if the desired long-term goals are to be realized. It is clear that achieving the short-term climate targets is already proving extremely difficult. In industrialized countries, domestic action has so far not produced very impressive reduction figures. And where trend breaks in emissions have been achieved, this has been the result of unexpected developments (German reunification) or other policy goals, such as the closure of British coal mines, which have not been directly related to climate change. A gap between desires and reality can also be clearly observed in The Netherlands (see Box 1.1).

Chapter 3 of this book evaluates the current societal support in The Netherlands for climate-change policies and the actual behaviour of various actors, such as businesses, citizens and local policymakers. It explores whether, and if so how, climate-change issues play a role in the decision-making behaviour of societal actors. If, over the coming decades, we are going to realize the long-term levels of stabilization as described above, the current situation will have to change. Chapter 3 concludes with seven dilemmas that actors in society may currently face (to differing degrees) when dealing with climate change.

An important issue in taking action is the variety of perceptions that exist in society about reducing emissions. At one extreme, there is the position that solving the climate problem will have major ramifications and that it will bring industrialized societies to the brink of economic collapse. At the other extreme, the position is that there are many solutions for dealing with the climate problem and that solving the problem is indeed affordable without major economic sacrifices - and there will even be benefits.

In this discussion, it is interesting to note that over the last 20 years, the final energy costs (as a percentage of GNP) for countries that are members of the Organization for Economic Co-operation and Development (OECD) have more than halved and are now in the order of 6-7 per cent of GNP (Blok, 2000). For The Netherlands, it has been calculated that although the costs for reducing CO_2 emissions will increase drastically in absolute terms, the relative costs will only rise slightly (but will stay below 2 per cent GNP) so long as the Dutch GNP continues to rise (Bezinningsgroep Energiebeleid, 2000). IPCC (2001c) comes up with similar numbers for the costs of reaching the Kyoto targets in industrialized countries, but it also points to the fact that lower stabilization levels will lead to higher costs. Since the share of energy in GNP as a whole apparently declines and the economic consequences of climate policies remain limited, one may conclude that this should offer opportunities for developing long-term policies on climate change

Box 1.1

CO_2 targets in policy documents in The Netherlands and international developments

Year	Target
1988	Toronto target: reduction of CO_2 emissions by 20% of 1988 levels by the year 2005 as an initial global goal.
1989	National Environmental Policy Plan (NEPP): short-term target of stabilization of CO_2 emissions in 2000, based on 1990 level; long-term target of stabilizing global emissions on a level that can be maintained by oceans and biosphere
1990	After the change in government, the new government presents a stricter version of the NEPP in the form of NEPP-plus: stabilization of CO_2 emissions by 1994-1995 and a new target for 2000, i.e., reduction of CO_2 emissions by 3% in 2000, compared to 1990 levels, and a possible reduction of CO_2 emissions by 5% in 2000, compared to 1990 levels, if international developments allow, and based on decisions made in 1995.
1990	The Netherlands states that together with other countries, they will research the Toronto target as an option for more drastic climate policy. It is clear that this will go beyond no-regret policies (Memorandum on Climate Change).
1993	Second National Environmental Policy Plan: reduce CO_2 by 3% in 2000, compared to 1990 levels, and if international developments allow (and based on decision making in 1995), a possible reduction of CO_2 emissions – by 5% in 2000, compared to 1990 levels (NEPP-II).
1995	Given the slow international progress and the difficulties of realizing reductions domestically, the government maintains its -3% CO_2 target but will not go for -5% (Letter to Parliament about CO_2).
1996	A goal of post-2000 stabilization of CO_2 emissions within The Netherlands of at least -3%, compared to 1990 levels, if international conditions allow (2nd Memorandum on Climate Change). Further reductions of greenhouse gases after 2000 by 1%-2% per year in industrialized countries.
1997	EU target of -15% and Dutch target of -10% in CO_2 emissions as part of EU burden-sharing agreement before Kyoto.
1997	At Kyoto, the introduction of a basket approach of 6 greenhouse gases, international flexibility through Joint Implementation, Clean Development Mechanisms and Emission Trading and inclusion of sinks. Worldwide reduction of –5.2% in period 2008-2012.
1998	Dutch target as part of the EU burden-sharing agreement: -6% between 2008-2012, compared to 1990 levels, and for the EU as a whole, -8% for all greenhouse gases.
1999	The Netherlands is one of the first industrialized countries to publish an implementation plan aimed at achieving its Kyoto targets (Memorandum on Implementation of Climate Change Policies).
2001	The Fourth National Environmental Policy Plan looks 30 years ahead and considers the transition towards a sustainable energy system crucial for dealing with the climate problem.
2001	At COP-6 bis in Bonn, a political agreement is reached about the rules of the Kyoto Protocol. This deal is made after earlier unsuccessful negotiations at COP-6 in The Hague (November 2000).
2001	At COP-7 in Marrakech, the COP-6 bis deal is finalized and put into legal language. This should make ratification of the Kyoto Protocol possible before the Earth Summit in Johannesburg in 2002.

Box 1.2

Some key figures about The Netherlands

- Population 2001: 16 million
- Number of households 2001: 6.9 million with a stronger increase in number of households than in population (2001: 2.4 million 1-person households)
- Population density is high: in 2001 472 inhabitants per square km on average, with about 1000 inhabitants per square km in the so-called Randstad (Western part of The Netherlands)
- The political culture is characterized as the 'Poldermodel', a culture that emphasizes dialogue and collaboration between different societal interests in developing new policies. Polder refers to the low parts of the country, which are vulnerable to flooding, areas that traditionally require a high level of collaboration to keep them safe for their inhabitants. This approach has been criticized for its alleged slow process of decision making.
- The economy of The Netherlands is very open. It is a net exporter. This is because of its geographical position and the port of Rotterdam, as well as large, well developed and well accessible neighbouring countries upstream of the main rivers. Industry and agriculture export a relatively large proportion of their production.
- The economic growth of The Netherlands GNP over the period 1990-2000 averaged 3.5% per year. Energy use in that period increased approximately 1% per year.
- In The Netherlands added value is mainly created in the trade, services and public sector (65.9%) and in industry (16.3%). Agriculture contributes with 2.4% and energy supply with 1.9%.
- Imports and exports in The Netherlands are still growing, resulting in an increase of transport of goods and services and a subsequent increase in the pressure on the environment.
- The emissions of all six greenhouse gases increased about 8% between 1990 and 1998. The target for the first budget period is to reduce emissions by 6%, compared to 1990 levels.

Source: RIVM (2001), RIVM/CBS (2001).

and for achieving targets. But even this seemingly favourable situation has not led to a reduction of emissions in The Netherlands over the last 10 years.

Achieving trend breaks: some historical examples

In light of the current inability to meet either the agreed-upon short-term targets for reducing emissions or even modest targets, it would appear that the long-term goals for drastic reductions in greenhouse gas emissions are extremely ambitious. If minor reductions are already proving difficult in the short term, will major transitions be achievable in the long run? It may be tempting to think of many relatively easy small steps (annual 1-2 per cent CO_2 reduction) leading to an 80 per cent reduction in 50 years. But the annual growth in the economy must also be considered, since it is directly linked to increasing emissions of greenhouse gases. Radical innovations are therefore necessary - innovations that will require changes in many domains in society.

In implementing the technologies necessary to achieve a climate-neutral society, government and businesses will face many obstacles because technologies will have to compete with established practices and technological trajectories. One should be aware that changes in the order of magnitude discussed here cannot result from central government directives. In the first place, in our modern Western societies, the span of control of national government is diminishing. In the second place, one also has to acknowledge that the type of transitions we discuss in this book are society-wide changes that to a large extent depend on a multitude of diffuse and independent decisions made by individuals, businesses and other societal institutions. Some speak of the *network society* (Castells, 1996: 243-307). It should be noted that these decision makers have not previously included energy, climate or the environment in their considerations and daily routines and have in many cases not been expected to do so. Achieving significant society-wide changes in greenhouse gas emissions over a period of several decades may imply that 'powerless states' have to take up a stronger role in this network society and its liberalized markets. But governments may no longer be able to exert the influence needed.

Will our modern society be able to deal with the ambitious targets that have been set? It may be tempting to say *no*, but targets that appear too ambitious should not put us off. History provides many examples of radical changes that were made in relatively short periods of time (some even within a few years or a few decades). Environmental policies set goals for drastic changes, including the need for major investments and ongoing innovations to make emission reductions feasible. In 1970, The Netherlands launched a 20-year programme to reduce water pollution, calling for a 60 per cent reduction in pollutants. The Netherlands seems to have a tradition of setting ambitious medium- and long-term targets. In 1989, the first National Environmental Policy Plan in The Netherlands contained a series of targets aimed at reducing a large number of emissions by 70 to 90 per cent within 20 years. More recently, in 1997, the Dutch Minister of the Environment stated that environmental impact per unit of consumption should be reduced by a factor of four in 25 years. Results from the third example can of course not yet be given, but the 60 per cent reduction in water pollution was achieved almost within the allotted time (Vermeulen et al., 1994). The second example refers to 73 industrial emissions to water and air, all with reduction targets in the range of 50–90 per cent for the period 1985–2000. By 1996, 51 of them had already been achieved. It appeared that another 11 would be achieved in 2000, but the remaining 11 emissions (including CO_2) turned out to be problematic (Ministerie van VROM and VNO/NCW, 1998).

An impressive example of rapid change can be found in the national energy infrastructure of The Netherlands. This transition demonstrates the need for simultaneous institutional and socio-cultural changes alongside technological developments. The Netherlands' energy infrastructure changed from mostly coal-based energy production to gas-based in the mid-1960s. This is an interesting example of a managed transition. It reveals powerful co-operation between business and state, carefully planned over a short period, implemented at the same overwhelming pace and building on the experience and expertise of a few actors in society. Within barely 10 years of the discovery of large supplies of natural gas in

Box 1.3

A short history of the 1960s gas revolution in The Netherlands

'Large scale transitions cannot be organized.' This statement no longer holds true after the gas revolution in The Netherlands in the 1960s. Within barely 10 years of the discovery of large supplies of natural gas, the Dutch energy infrastructure was completely turned around.

The transition was impressive. The change households were just starting to make from coal-based to oil-based heating was completely overtaken by the transition to natural gas. Within five years the market for domestic heating appliances had completely converted to gas-based heating. In 1963, only 10 per cent of all new dwellings had central heating. Five years later it was 80 per cent. All 5 million domestic gas stoves (fuelled by gas from local gas factories) had to be converted or replaced. This involved 5000 different types of appliances. In an operation of almost military precision, 1.7 million appliances were replaced by new ones and 3.3 million were converted. Figure 1.3 shows the rapid transition in the market for new cooking appliances.

During the 1950s, the Dutch economy and households were largely fuelled with coal, oil and gas from local gas factories using coal and some peat. In 1948-1950, the first natural gas supplies were discovered in the north of The Netherlands by NAM, an oil extraction company jointly owned by Shell and Esso. At first they were not at all interested: gas was considered an annoying by-product of oil extraction and was sold to local gas companies. The situation changed drastically after tremendous new discoveries in 1959. NAM initially kept it silent for more than a year, until a Belgian member of the European parliament revealed it in October 1960.

This was the start of a very rapid chain of events. The market parties involved realized they could earn substantial revenues from these gas reserves, as could the Dutch government. As a result, a large-scale transition was made in the traditional energy infrastructure to a gas-dominated infrastructure.

A central planning approach was used, with close co-operation between a small number of large market actors and the national government. This planning approach included re-institutionalization, orchestration of the phasing out of coal stoves and conversion of gas stoves, establishing prices and addressing socio-economic side effects.

The high pace of the transition was remarkable. Natural gas was discovered in the country in July 1959, but it was only made public in October 1960. Esso completed its major implementation plan in December 1960 and secured agreement on it with its NAM partners in March 1961, before submitting it to the Dutch Minister of Economic Affairs. The subsequent NAM-government negotiations lasted until mid-1962. In July 1962, a mem-

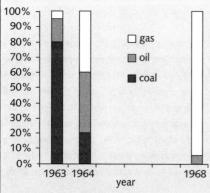

Figure 1.3 *Transition in sales of stoves*

Box 1.3 *continued*

orandum was sent to the Dutch parliament. It was accepted in October 1962. The transition to natural gas required a national-level institutional structure, in contrast to the existing, locally oriented institutions. In April 1963, the national government established a national organization, the Gasunie, as the nationwide distributor. From that moment on, implementation could start, resulting in a complete transition within five years.

Initially, the Esso parent company in the USA did not recognize the opportunities. However, in 1960, they saw the situation in The Netherlands as a chance to reduce the role of the pipeline companies, which was something they had just missed in the USA. Getting a foothold in this market would only be possible if the enormous costs of establishing the pipeline infrastructure could be covered. That required complete and rapid conversion of all Dutch households to heating and cooking with natural gas. Esso's partner, Royal Shell, first reacted sceptically, but soon followed suit. In 1962, Esso issued a report that identified many institutional, socio-economic, social and psychological barriers.

Institutional barriers

Gas supply was organized at the local level. This was the start of inter-municipal co-operation. Provinces were trying to assume a co-ordinating role in the local gas markets. After the Esso report, the national government intervened with the establishment of the Gasunie.

Prices were not set in a free market; they were artificially set through negotiations amongst the tradition local gas producers, national government and Esso and Shell. Market prices for traditional energy resources served as a guide. This resulted in a regressive price system, allowing financial stimuli for conversion and replacement of stoves and heating systems. Distributors paid 20 per cent of the conversion costs. Replacement allowances were high.

Technical barriers

There were technical barriers to be addressed: all gas appliances had to be converted because the natural gas had a higher caloric value than the manufactured gas. Some municipal gas companies, which were in charge of certifying appliances, set requirements that were too stringent, thus hindering the diffusion of new appliances.

Social barriers

Of course such a change in the energy infrastructure raised public resistance. An entire sector – mostly small local enterprises in the coal and oil retail sector – would lose their market. Some of them were offered jobs in the conversion programmes or in other programmes. Fitters also resisted, partly due to a lack of knowledge, but also because installing new gas appliances was more expensive, giving them smaller margins. At first, small-scale industries didn't convert because of unfavourable costs; however, this changed after 1970 when their gas price was no longer linked to the price of fuel oil.

The transition was accompanied with a marketing campaign that was impressive for the time. All kinds of information campaigns were used: stickpins, village parties at the moment the village was connected to the natural gas grid, advertising in newspapers and, after 1968, also on television. Selling points such as comfort, health and logistic advantages were intensively communicated.

Source: Schot, J.W. et al. (eds.), 2000.

the country, the Dutch energy infrastructure was completely turned around. Yet it is doubtful whether a comparable programme could be implemented today (see Box 1.3).

The cleaning up of Dutch waters mentioned earlier also involved comparable institutional changes. Reducing water pollution required the development of new technologies, but even more, it required institutional changes within government agencies. The existing structures and institutions were re-shaped during this process from small-scale organizations with a single objective, strongly oriented towards agricultural needs, to multi-purpose, more democratic organizations.

A third historical example also illustrates the possibilities and mechanisms of socio-cultural change. Smoking behaviour has changed considerably in many Western countries. In The Netherlands, the transition mainly took place between 1970 and 1988, when the number of smokers declined from 60 per cent of the population to 33 per cent, a level that has remained stable since then. In addition to increased taxation, the government used information campaigns stressing the effect of smoking on health. In her study about 'the status of voluntary moderation', Aarts (1998) described this as an example of the 'trickle-down effect' where high-status groups may adopt forms of self-restraint and voluntary moderation as a way of distinguishing themselves from other groups in society. These other groups may then follow the behaviour of the higher-status groups (Aarts, 1998: 73, 282). One might expect such mechanisms to be helpful in changing culture, but Aarts is rather pessimistic about this because essential features for far-reaching cultural change are absent (Aarts, 1998).

One lesson to be learned from these examples is that in elaborating the possible paths towards a drastic reduction of greenhouse gas emissions, the establishment of technological scenarios is merely a first step. Trend breaks and transitions are always a co-ordinated development of changes in technology, the economy and society (De Vries, 1994: 296; Jansen, 1994: 499). Most of the chapters in this book address developments in society and focus on structural, institutional and cultural barriers to achieving a climate-neutral society, along with the co-evolution of technology and society to reach this goal.

Using the concept of trends, trend breaks and transition management

In the previous sections, we used the term 'trend break' and we will continue to do so, but it requires some explanation, as does the use of the terms 'transitions' and 'transition management'. This vocabulary springs from Dutch literature on long-term environmental policy-making (see Box 1.4). It is useful to elucidate some of the discussions on these concepts and the theory of 'trend breaks' and transitions.

In order to realize drastic reductions in CO_2 emissions, societal trend breaks will take place against the background of dominant, almost autonomous, societal developments, such as individualization, demographic trends, internationalization and rapid technological developments (especially in information and communication technology and services) (see Box 1.5). Technological innovations may allow for reduced levels of CO_2 emissions, but one also has to take into account the existence of rebound effects. This is the phenomenon of increased consumption of a service (e.g., lighting or transportation) when the variable cost for that service

Box 1.4
Definitions

In this book, we use, rather loosely, a number of closely related concepts such as *trends, trend breaks* (shifts in trends), *transitions, innovations* and *transition management.*

Trend is defined as a gradual change in a particular observed variable, which originates from other, possibly interrelated, developments in society, is constructed from human actions, persists for a long time and covers a certain domain (Slob et al., 1999).

The term *trend break* is not used in English. It is a direct translation of the Dutch word *trendbreuk. Breuk* can be interpreted as 'discontinuity' and is closely connected to the idea of 'shifts in trends' or 'radical changes' (De Vries, 1994). A *trend break*, therefore, is a change in a current trend that persists for a long period of time and covers a certain domain. An appealing element in the Dutch meaning of *trend break* is the sense of urgency it implies, which we consider relevant in the context of climate change.

Transitions are defined by Rotmans et al. (2000) as long-term, gradual, continuous processes of structural societal change. Transitions take place through mutually reinforcing and counteracting developments in technological, economic, ecological and socio-cultural domains. Transitions, according to Rotmans et al., should be distinguished from trend breaks and *innovations*, which they consider to be developments in a specific domain that can contribute to the transition process. In their view, the transformation of the current society into a climate-neutral society can be defined as a *transition process*, because it implies a system *innovation* (as opposed to optimization). The transition to a climate-neutral society can also be seen as a collection of breaks with current trends. *Transition management* is an approach designed to deliver an active contribution to the shaping of transitions.

decreases following an attempt to reduce energy use/emissions through an investment (such as a low-energy lamp or a fuel-efficient heater).

The concept of trend breaks will be used in two ways throughout this book. Trend breaks can be used first of all in a *normative way* to express necessary changes. In this book, the trend breaks required to achieve a climate-neutral society are identified. And second, the concept of trend breaks can be used to *analyse constraints and opportunities* in the transition towards a climate-neutral society. The emphasis in this book will be on this second way of using the concept. We argue with De Vries that it would be rather naive to think that we can 'break' trends if they go against underlying dynamic forces (De Vries, 1994: 295-303). Provoking trend breaks implies identifying such underlying forces and managing simultaneous changes in technology, culture and social and institutional structures to shift the direction of societal developments.

To explore new directions for society (such as the transition to a climate-neutral society) a method called 'back-casting' can be applied. 'Back-casting' can be defined as a participatory method that explores the innovations necessary to realizing a collectively formulated picture of a sustainable future in the long term (Jansen, 1994: 502; Weaver et al., 2000: 121; Jansen, 2000: 173-180). Or, as Robinson (1990) states, 'the major distinguishing characteristic of backcasting

Box 1.5

Examples of dominant societal trends

	Most important trends	Positive environmental effect	Negative environmental effect
NL in international economy	Large, stable share of EU in international trade Growing importance of transport and services	Services have relatively small impact	Increasing emissions due to transport
Technology	In some fields, NL is behind in R&D and new technology		Delayed upgrading: insufficient decrease in environmental pressure per unit of added value
Demography	Increasing age of population Increasing number of immigrants More and smaller households		Higher energy consumption due to smaller households
Lifestyle and preferences	Individualization Risk avoidance Increasing ownership of durable commodities Increasing mobility Housing preferences: suburban or city centres, one-family dwellings Increasing leisure time spent abroad	Possibly green consumption pattern Increasing use of space for housing and traffic	Strong growth in house-hold use of energy and materials
Institution/ government role	Liberalization New forms of business-government co-operation Increasing role of EU	Phasing out of support schemes for certain hazardous activities Increased efficiency in use of energy and resources caused by market stimuli	Focus on short-term revenues Loss of direct influence on relevant decisions in former public sector
Labour market	Shortage of educated personnel Increase of flexible work Increase of part-time work Rising levels of education Increasing participation of women Concentration in centre, west and south of NL	Increased working at home (less traffic)	More use of space and appliances per labourer

Source: Felsö et al. (1999), Nijkamp en Verbruggen (2001), Slob et al. (1999).

analysis is a concern, not with what futures are likely to happen, but how a desirable future can be attained.'

In this approach, back-casting always includes focusing on opportunities and constraints. In this book, images of possible futures (such as those described in Chapter 2) are used in identifying barriers and opportunities for innovation and diffusion, relevant interests, roles and responsibilities, and economic and administrative institutions that will need to be reshaped in order to achieve the long-term target of an 80 per cent reduction in the emission of greenhouse gases.

The next question that arises is how to realize the necessary trend breaks and orchestrate the transition towards a climate-neutral society. In the context of the societal trend breaks necessary for successful development and implementation of the technologies required for a climate-neutral society, it would be wise to open the window of opportunity as wide as possible (Rotmans et al., 2000: 71). In a recent contribution to the debate on long-term policies, Rotmans et al. apply the concept of 'transition management', stating that it should be characterized by the following:

- long-term perspectives as a framework for assessing short-term policy;
- reasoning in terms of multiple domains and various actors;
- governance based on learning processes;
- aiming at system innovation and improvements;
- keeping ranges of opportunities open (a broad playing field).

They predict that a drastic transition in the energy system will take at least two generations, based partly on a retrospective analysis of historical transitions from wind to steam (in the 19th century) and from coal to electricity (in the early 20th century). In Chapter 4, possible ways to orchestrate this transition will be analysed and discussed in detail.

The structure of this book

Chapter 2 presents two long-term visions of a climate-neutral energy system in The Netherlands. It looks at the window of opportunity and shows that there are a number of technically feasible ways to realize drastic emission reductions. At the same time, it is well known that climate change is a tough problem to deal with and that the results of current policies have been relatively meagre. Chapter 3 goes on to identify a number of dilemmas that play a role in current climate-change policies. These dilemmas are elaborated in the light of desired long-term developments. Chapters 2 and 3 together illustrate a sharp contrast between the problematic current situation and the substantive targets that need to be achieved over time. The question, then, is how to realize trend breaks and orchestrate the transition towards a climate-neutral societal. Chapter 4 looks at the governance of technological change and innovation.

Chapters 2, 3, and 4 together form the basis on which the authors of the subsequent chapters have written their analysis about the transition towards a climate-neutral society. As a whole, the book provides the reader with an analysis of prospects of long-term climate policies, short-term issues, the societal trend breaks needed to realize substantive emission reductions and possible ways to realize these trend breaks. However, this book can't provide a complete picture of all relevant elements of such a transition. There are many relevant issues that are beyond the scope of this book, such as the role of business and other new technologies such as genetic engineering, to mention just two.

Chapter 5 analyses households, behaviour and consumption patterns. It assesses future opportunities for drastic changes that would lead to more environmentally sound energy consumption by households and discusses constraints to changes in behaviour. In Chapter 6, the role of local authorities in the transition to a climate-neutral society is discussed. Chapter 7 takes the perspective of materials management as a breakthrough technology for reducing greenhouse gas emissions. In Chapter 8, the contribution that information and communication technology (ICT) (as another breakthrough technology) can make to the transition is discussed and some of the pitfalls related to ICT are identified. Chapters 9 and 10 then take more of a governance perspective. In Chapter 9 the economic preconditions for drastic emission reductions are examined, along with the possibilities of emission trading. Two forms of trading are analysed from a policy lock-in perspective. In Chapter 10 the legal possibilities of achieving long-term reductions are analysed. Chapter 11 presents a stakeholder perspective on long-term reductions, based on the results of an 18-month dialogue in The Netherlands about drastic emission reductions in the long term. The visions presented in Chapter 2 form the starting point for this dialogue. Chapter 12 is a concluding chapter by the editors.

References

Aarts, W., *De Status van soberheid*, Academisch proefschrift Universiteit van Amsterdam, Amsterdam, 1999.

Berk M.M. et al., *Keeping our options open. A strategic vision on near-term implications of long-term climate policy options*, NRP Program Office, Bilthoven, 2001.

Bezinningsgroep Energiebeleid, *Klimaatprobleem. Oplossing in zicht*, Delft, 2000.

Blok, K., *Energie in de 21ste eeuw. Technologische en maatschappelijke uitdagingen*, Oratie Universiteit Utrecht, 2000.

Castells, M. *The Information Age: Economy, Society and Culture Volume I: The Rise of the Network Society*, Blackwell Publishers, Massachusetts, 1996.

De Vries, H.J.M., Trends and discontinuities: their relevance for sustainable development In: Dutch Committee for Long-term Environmental Policy, *The Environment: Towards a Sustainable Future*, Kluwer, Dordrecht, 1994, pp. 277-310.

Felsö, F. et al., *Milieurelevante trends in de Nederlandse samenleving*, Publicatiereeks milieustrategie 1999/6, Den Haag, 1999.

Intergovernmental Panel on Climate Change (IPCC): Nakicenovic, N. et al., *The IPCC Special Report on Emission Scenarios*. Intergovernmental Panel on Climate Change (IPCC), Cambridge University Press, U.K./U.S.A, 2000.

Intergovernmental Panel on Climate Change (IPCC): Houghton, J.T., Y. Ding, D.J. Griggs, M. Noguer, P.J. van der Linden, X. Dai, K. Maskell, C.A. Johnson (eds.), *Climate change 2001. The scientific basis.* IPCC Working Group I, Cambridge University Press, U.K./U.S.A, 2001a.

Intergovernmental Panel on Climate Change (IPCC): McCarthy, J., O.F. Canziani, N.A. Leary, D.J. Dokken, K.S. White, *Climate change 2001. Impacts, Adaptation, and Vulnerability*, IPCC Working Group II, Cambridge University Press, U.K./U.S.A., 2001b.

Intergovernmental Panel on Climate Change (IPCC): B. Metz, O. Davidson, R. Swart, J. Pan (eds.), *Climate change 2001. Mitigation.* IPCC Working Group III, Cambridge University Press, U.K./U.S.A., 2001c.

Jansen, J.L.A., Towards a sustainable future: en route with technology! In: Dutch Committee for Long-term Environmental Policy, *The Environment: Towards a Sustainable Future*, Kluwer, Dordrecht, 1994, pp. 497-523

Jansen, J.L.A. et al., On the search for ecojumps in technology: from future visions to technology program. In: Thompson Klein, J. W. Grossenbacher-Mansuy, R. Häberli, A. Bill (eds.), *Transdisciplinarity: Joint Problem Solving Among Science, Technology, and Society*, Birkhauser Verlag AG; Basel, 2000, pp.173 –180.

Kok, M.T.J. et al., *Klimaatverandering, een aanhoudende zorg*, NOP Programmabureau, NOP rapportnr. 410 200 113, Bilthoven, 2001.

Ministerie van VROM/VNO-NCW, *De stille revolutie: industrie en overheid werken samen aan een beter milieu*, Distributiecentrum VROM, 1998.

Nijkamp, P. en H. Verbruggen, Global trends and climate change policies, in: E. van Ierland, J. Gupta and M.T.J. Kok (eds.), *Options for international climate policy*, E. Elgar, Cheltenham, 2001.

RIVM, *Milieubalans 2001. Het Nederlandse milieu verklaard*, Kluwer, Alphen aan de Rijn, 2001.

RIVM/CBS, *Milieucompendium 2001*, Kluwer, Alphen aan de Rijn, 2001.

Robinson, J., Futures under glass: a recipe for people who hate to predict, *Futures*, vol. 22, October, 1990.

Rotmans, J., R. Kemp et al., *Transities en transitiemanagement. De casus van een emissiearme energiehuishouding*, ICIS, Maastricht, 2000.

Slob, A.F.L. and Th.M.M. van Hoorn, *Major shifts in societal trends and their impact on climate change*, NRP report nr. 410200012, Bilthoven, 1999.

Schot, J.W. et al (eds.), *Techniek in Nederland in de 20ᵉ eeuw. Deel II Delfstoffen, energie, chemie*, Walburg Pers, Zutphen, 2000

UNDP, *World Energy Assessment*, UNDP/UNDESA/World Energy Council, 2000.

Vermeulen, W.J.V., C. Dieperink, P. Glasbergen, Towards an Integral Water Management: Changing Water Quality Management in The Netherlands, In: Löwgren, M. and R. Hjorth (eds.), *Environmental Policies in Poland, Sweden and The Netherlands: a Comparative Study of Surface Water Pollution Control*, Linköping, 1994, pp. 15-45.

Weaver, P., et al., *Sustainable Technology Development*, Greenleaf Publishing Ltd, Murfreesboro, 2000.

Chapter 2

Transforming the energy system of The Netherlands

Two visions on reaching 80 per cent emissions reduction by 2050

André Faaij [1]

In this chapter...

... two long-term visions of the Dutch energy system of the future will be presented, analysed and evaluated. Both visions were formulated with the aim of meeting the requirement of reducing greenhouse gas emissions by 80 per cent compared to 1990 levels. The two visions adopt very different perspectives and contexts and differ significantly in the way these emission reductions will be achieved. The two, rather opposing, visions show that it is feasible to combine (strong) economic growth with far-reaching reductions of greenhouse gas emissions provided a number of key options are developed and applied. It will be shown that drastic changes in many areas of society are required to reach the desired greenhouse gas emission levels in 2050. However, the results also show that this objective can be realized in different ways. In other words, there are various ways of achieving a low greenhouse gas emission economy in the long term.

Introduction

Various long-term projections for the world economy suggest that economic growth is likely to remain strong for many decades. Increased consumption and industrial production will lead to substantial increases in greenhouse gas emissions if current technologies and fossil fuels remain the basis of the energy and production systems. This chapter explores the possibilities of combining steady economic growth with a stringent objective: reducing greenhouse gas emissions in 2050 by 80 per cent compared with 1990 levels. We consider all the main possibilities for reducing greenhouse gas emissions, including increased efficiency in energy and material use, renewables and advanced fossil fuel options, which could play a role within the time-frame considered. This evaluation covers The Netherlands, a country with a high population density and high per capita energy use.

In 1998, The Netherlands emitted about 240 Mton CO_2 equivalent. The total emissions can be divided into two groups The first is made up of non-CO_2 greenhouse gases such as methane (CH_4), N_2O (a greenhouse gas released primarily

---

[1] This chapter is to a large extent based on the following publication: A. Faaij, S. Bos, J. Oude Lohuis, D. Treffers, C. Battjes, J. Spakman, R. Folkert, C. Hendriks, *Sustainable energy systems in the long term - two visions of the Dutch energy system*; a report prepared for the COOL dialogue by the Department of Science, Technology and Society, Utrecht University, Netherlands Energy Research Foundation, National Institute for Public Health and Environment and Ecofys., October 1999. This exercise was originally conducted to provide a starting point for the COOL dialogue in The Netherlands, in which a wide range of stakeholders conducted a structured dialogue on long-term climate policies. The COOL project is discussed at length in Chapter 11 of this book.

during the use and production of artificial fertilizers), and CFCs, which together account for about 23 per cent of the warming potential. The remaining 77 per cent are CO_2 emissions, almost exclusively caused by the combustion of fossil fuels to produce electricity and for industrial processes, transport and domestic applications, such as heating. The total energy consumption in 2000 amounted to about 3000 PJ. Over 97 per cent of this energy use is covered by fossil fuels. Renewable energy (solar, wind, biomass, hydro, geothermal) plays a very modest role in the energy supply and covers just 1 per cent (35 PJ, of which 30PJ from biomass and waste) of the energy demand. The remaining 2 per cent of demand is met by imports of nuclear energy and electricity.

The Netherlands is one of the richer countries within the Oragnisation for Economic Co-operation and Devlopment, with a GDP of about US$ 365 billion (1999). The energy intensity (energy use per unit of GDP generated) is relatively high compared to other countries. This can be explained by the relatively large share of energy-intensive economic activities, such as heavy industry (for instance, the chemical industry), greenhouse horticulture and transport, in the economy. The Netherlands accounts for roughly 1 per cent of both the world's GDP of approximately US$ 40,000 billion (1999) and the global energy use of about 400 EJ (400,000 PJ).

Furthermore, all these economic activities are concentrated in a small area: The Netherlands covers 3.5 million hectares of land (compared with a global surface area of about 13.2 gigahectares or 13,200 million hectares), or less than 0.03 per cent of the world's land area. Consequently, with its 16 million inhabitants (0.27 per cent of the world's population), The Netherlands is one of the most densely populated countries in the world.

Economic growth has been steady and relatively high over the past decade. Annual growth figures have fluctuated between 2 per cent and over 4 per cent in this period. Although energy efficiency has also gradually improved (as evidenced by lower energy intensities for a wide variety of sectors), the total energy consumption of The Netherlands is rising and there has been no evidence of the decoupling of energy use and economic growth.

Measures to reduce greenhouse gas emissions have not led to a decline in those emissions. Non-CO_2 greenhouse gas emissions are declining, however, mainly because of successful policies targeting industrial emissions and a shift from CFCs to other compounds for various applications. However, this effect is more than offset by rapidly growing CO_2 emissions, mainly due to growing consumption of mineral oil (for transport) and natural gas.

The fact that economic growth has up to now invariably led to higher energy consumption is a worrying starting point, given that economic growth is a worldwide ambition and one that is being achieved. The world economy as a whole has recorded growth rates of 1-3 per cent over the past decade, with much higher figures for countries like China and India. Economic growth is expected to remain high in many developing regions. The world economy could develop and grow at an annual rate of 2-4 per cent. Such figures would lead to an expansion of global GDP by a factor of 4-9 in 2050! Without drastic changes in the way that energy is used and produced, such economic growth will also inevitably lead to a dramatic increase in CO_2 emissions to levels which are without doubt unsustainable for the world's climate.

The IPCC has argued that the risk of human-induced climate change can only be limited to acceptable levels if global greenhouse gas emissions are reduced to around 50 per cent of their 1990 levels (IPCC, 2000) in the course of the 21st century. This is in itself a very ambitious target given the current situation, but it becomes a daunting one if at the same time the world economy develops strongly. The IPCC has also produced a wide variety of scenarios regarding greenhouse gas emissions and the underlying economic and technological developments. While some of those scenarios lead to the desired emission reductions, others give rise to an almost uncontrollable growth of global emissions.

A key principle to emerge from the IPCC exercises is that the industrialized countries have been responsible for the bulk of greenhouse gas emissions in the past and the subsequent build-up of greenhouse gas concentrations in the atmosphere. Developing countries that desperately need economic growth simply to meet basic standards of living should therefore be allowed a larger share of the tolerable emission volume. Consequently a relatively large share of greenhouse gas emission reductions would have to be realized by industrialized countries. Following this rationale, an 80 per cent reduction in greenhouse gas emissions by those countries is often mentioned as a desirable goal for this century.

The key objective of this chapter is to outline a number of quite different ways in which a prosperous industrialized country like The Netherlands could meet this target of reducing its greenhouse gas emissions by 80 per cent by the middle of the 21st century (2050). These scenarios, or 'visions', of the Dutch energy system serve as a point of departure for exploring where in society major transitions are required in order to reach the 80 per cent emission-reduction objective and whether they are feasible or likely. In the course of the discussion in this chapter we will highlight some types of measure and transition that could play an important role. Examples include changes in behaviour and consumption, structural changes in the economy, greater energy and material efficiency, the potential role of renewable energy sources, the use of CO_2 storage when using fossil fuels and measures to reduce greenhouse gases other than CO_2.

The two visions of the Dutch energy system that can result in 80 per cent reduction of greenhouse gas emissions are constructed as follows:

- on the basis of two recent IPCC scenarios (A2 and B1) for the world, we have composed two quantitative descriptions of the Dutch economy in 2050. Those descriptions include estimates of parameters at a sectoral level, such as physical production by the major industries, transport kilometres, number of dwellings, etc.
- we have made an inventory and evaluation of options to reduce greenhouse gas that could be available in the long term. This inventory includes options for improving energy efficiency in industry, the built-up environment and transport, a multitude of energy-supply options (renewables, biofuels, nuclear energy, advanced fossil fuel technologies with CO_2 sequestration), as well as options that lead to reduction of non-CO_2 greenhouse gases.
- linked to the two scenarios and the principal socio-economic conditions assumed in those two different 'worlds', sets and combinations of measures have been chosen that are likely to play a role in meeting the far-reaching emission-reduction target. For example, renewables play a key role in one vision, hydrogen infrastruc-

ture and CO_2 storage a key role in the other.

- using a quantitative analysis of the selected technology mixes, we have calculated the resulting net greenhouse gas emissions and determined the energy balances for The Netherlands for both visions.

This chapter concludes with a list and summary of topics that are crucial if the radical change in greenhouse gas emissions is to be achieved. Those topics are the key subjects for the subsequent chapters in this book and include aspects such as local and international policy-making, household behaviour and technological innovation. In addition, by presenting two quite different visions for reaching the same objective, this chapter will also provide a basis for identifying major barriers that already exist or can be expected in the future to achieve the radical changes that are required in society. This chapter basically 'sets the agenda' for the authors who will later discuss the various issues in more detail by identifying and discussing the possibilities, barriers, limitations and essential actions relating to the major transitions needed to meet the 80 per cent reduction target.

Two visions of The Netherlands in 2050

It is possible to distinguish several key drivers that determine the level of a society's greenhouse gas emissions. The first major factor is the pace of economic growth, which is directly related to the social and cultural values of a society and the level of consumption and wealth that is considered desirable. A further distinction can be made between a society that gives priority to individual freedom and a more socially-oriented society. Another distinction can be made between a world economy structured on global competition and open markets and an economy with greater government control and oriented towards (regional) trading blocs. Other key variables are the way in which energy and resources are used (in other words, how efficiently they are utilized) and the type of resources (like energy carriers) that are used.

Logically, all those drivers are mutually influential. The predominant value pattern and philosophy is likely to lead to specific preferences for technologies, or perhaps it is better to say that the technologies and resources used are derived indirectly from cultural values.

Table 2.1 gives a summary of the characteristics of two alternative potential futures and different contexts for The Netherlands in 2050. Of course, these combinations of characteristics are in many respects arbitrarily chosen. On the other hand, the factors are combined in such a way as to produce consistent visions.

These 'visions' merely serve as examples to illustrate how a target of 80 per cent reduction of emissions could be achieved in the potentially very different worlds in which it may have to be realized.

The selected visions differ considerably in their assumptions with regard to the situation in The Netherlands in 2050 (behaviour, population, economic and spatial structure, etc.), without actually being 'extreme'. The visions sketch a picture of a conceivable situation in The Netherlands in 2050. The two visions will be referred to simply as 'A' and 'B'.

Table 2.1 *Qualitative differences between the visions*

	Vision A	Vision B
World view	Internationally-oriented "Global Village" World-wide convergence	Regionally-oriented, world trade blocks
Social environment	Individualistic	Sociable, family-oriented
	Little appreciation of environment and nature	Environmentally-minded
	Personal interest first	Distribution of wealth, social equity
	Well-being of fellow man subordinate to personal interests	Well-being of fellow man important
Economy & consumption	High economic growth and dynamic	Less dynamic and economic growth
	Part of world economy	Part of EU trade bloc
	Motivated by market mechanisms, little government intervention	Regulation, strong government
	Recreation abroad	More nature areas / recreation within The Netherlands
	Quantity above durability	Durable goods, quality
Use of space	Suburbanization	Careful use of space
	Nature areas fragmented	Large continuous nature areas
Traffic and transport	High demand for mobility in passenger and freight transport	High demand for mobility in passenger transport, moderate for freight transport
	A lot of private traffic	Public transport
	Road transport dominant	More freight trains and inland shipping
Energy & environment	Rapid diffusion of technology	Technology develops less quickly
	New infrastructures implemented	Possible solutions on demand-side important
	No change in behaviour	Willingness to adapt behaviour
	Cost-benefit analysis	Environment and sustainability given priority

Vision A

International relations

In this vision, the gap in prosperity between rich and poor countries has disappeared, partly due to rapid developments in transport and communication technology. International co-operation promotes productivity growth and facilitates the faster diffusion of technology. This has led to significant mobility of people and ideas.

The world is characterized by rapid and successful economic development. The most important factor behind the economic dynamism is the trust placed in market-oriented solutions. The level of investment is high, both in technology and in education. The emphasis in this vision is on market mechanisms and increasing productivity, partly thanks to the faster introduction of new and more efficient technology.

Social environment and demographics

People focus on personal progress and want the freedom to control their personal development. Ambition is important: people want to make financial progress and achieve something in their lives. The average standard of education is high. It is important to enjoy life and the level of consumption is high. Concern for nature, the environment and the well-being of others does not have priority. Individualism is the key characteristic of this vision. The workforce includes women as well as men and they work over forty hours a week.

Day care centres are commercial, as is the social service sector. Education is provided mainly by private schools. Steering and policy in society are dominated by technological, market-oriented solutions and approaches. Decision-making processes are fast and there are limited opportunities for public participation.

The family does not have a prominent role in society. The proceeds of the economic growth are used mainly to create further growth, and only to a small extent for social services and the environment and nature. In a world where the differences in prosperity between countries have disappeared, there are still groups within the regions that profit less from the prosperity. Differences in income widen and certain groups fall behind socially. In the multicultural society polarization has therefore taken place. Crime is dealt with repressively.

The world population has grown to nine billion people in 2050. The average age is higher than in 1990. In The Netherlands, the ageing of the population is quite advanced. The Dutch population, at 16.1 million inhabitants, is a little higher than in 1990. The high prosperity results in a long life expectancy and low birth and death rates. In combination with the importance attached to personal development, this has led to a large proportion of single-person and two-person households. The thinning out of the family is evident: The Netherlands has 10 per cent more inhabitants than in 1990 but there are 50 per cent more homes.

Economy and consumption

Due to the dynamics of the free market and the high mobility of people, goods and technology, the world economy (added value) is almost nine times larger in 2050 than in 1990. Although economic growth in the Organisation for Economic Co-operation and Development countries has been slower, total GDP in The Netherlands is over five times larger than it was in 1990. This implies the same pace of growth as in the last 50 years (including the powerful growth during the post-war reconstruction).

The income per head of the population in The Netherlands in 2050 is five times higher than in 1990. The per capita consumption of meat and dairy products is a little higher than in 1990. The most significant changes in spending patterns compared with 1990 are in foreign tourism and the buying of services. People go on foreign holidays several times a year. The aeroplane is by far the most popular form of transport for holidays; there are 15 times as many flights as in 1990. Many day trips and recreational activities are enjoyed within The Netherlands.

The car-ownership rate is high, there is extensive sub-urbanization and dense transport networks, both national and international. Most passenger and freight transport in The Netherlands takes place by car and truck. The number of kilometres driven for private purposes is 80 per cent higher than in 1990, and freight transport by road is a factor of 8 higher. Freight transport by rail is growing, but passenger transport by rail is lower than in 1990. The volume of inland shipping is roughly one-and-a-half times what it was in 1990.

Energy and the environment

In this world, possible problems that arise are solved on the basis of a cost-benefit analysis, in which the relevant aspects are assigned a (capital) value. In the process, aspects relating to nature, environment and durability are assigned a low value.

Energy and materials are abundantly available world-wide. This is mainly due to the rapid technical improvements in energy extraction. The energy prices are relatively low and stable so there is little incentive to save energy in this free market. Due to the pace of technological modernization, the prices of solar and wind energy are significantly lower than in 1990 but they still cannot compete with fossil-fuel energy on a large scale. Energy use is high and there is little willingness to pay higher costs for sustainable energy or energy-saving technology.

People are not prepared to adapt their behaviour in order to reduce CO_2 emissions; any measures that have to be taken must have minimal effect on material prosperity and the consumption pattern. People are prepared to pay for measures which allow them to maintain their pattern of consumption. This means that in this world technological solutions are quickly adopted, stimulated by market forces and powerful R&D. Due to the substantial use of fossil fuels, greenhouse gas emissions are mainly reduced by CO_2 storage and the import of (sustainable) energy carriers. Alternatives which fit into this vision are the use of nuclear energy and hydrogen technology. The most important factor behind the adoption of measures to increase material efficiency is the avoidance of the cost of the waste treatment.

Space

People desire comfort and space. Cities are therefore made up of large suburbs where people live in spacious park-like built-up areas. One-third more space is used for living than in 1990, although the population is only 10 per cent larger. Many people in the rapidly growing service sector work at home so that the space required for offices is the same as in 1990. Large industrial complexes take up around 30 per cent more space than in 1990. The Netherlands has several large airports due to the demand for air travel, which is 15 times higher than in 1990. There is a high degree of car ownership and the majority of freight transport is by road. The capacity of the road system is calculated to meet this demand. Broad motorways connect the city suburbs, industrial areas, ports and airports.

Due to the high price of land in The Netherlands the competitive position of the land-bound agriculture has deteriorated. Agricultural acreage has declined by a third in 2050 compared with 1990. Nature areas are exploited to meet the huge demand for recreation. The landscape is characterized by a park-like layout.

Vision B

International relations

In vision B, different regions in the world have developed in different directions. There is less international trade and interaction. There are also fewer powerful international institutions. In vision B, the emphasis is on regional identity and self-sufficiency. These important values are developed within the individual cultural and/or economic regions.

Social environment and demographics

Nature is highly valued. The well-being of people and animals and a good environment are regarded as just as important as prosperity. Social aspects play a prominent role in decision-making; people show great concern for others. Attention is given to distribution of wealth and equality in society. The classical family is just as important as in 1990. Men and women participate equally in the labour process and share responsibilities in raising their children. Day care centres and social services are part of the public sector.

The standard of education is high. The government runs education. People are doing well financially and want to contribute to the progress of society as a whole. Personal development is not of primary importance but goes hand in hand with the development of a sustainable and sociable society. People attach importance to quality and sustainability, quantity is less important. Society is structured in such a way that as many people as possible can be involved in decisions. Most of the proceeds of the economic growth are invested in a sustainable economy, social affairs and the environment and nature. Society in The Netherlands is multicultural. Immigrants make up a large part of the working population and are well integrated at all levels of society. Teaching standards and values at local level prevent crime.

In vision B, the world population in 2050 is 9.4 billion. The population of the Organisation for Economic Co-operation and Development as a whole shows a more moderate growth at 13 per cent than the world average. The Netherlands is an exception: in 2050 the country has 18.9 million inhabitants and has grown some 25 per cent compared to 1990. This is due to both the family-oriented nature of society and the results of an open immigration policy. The Netherlands in 2050 has considerably more elderly people than in 1990. The number of homes in The Netherlands has risen by some 15 per cent. The thinning out of households increases slightly in this vision.

Economy and consumption

In vision B, the world economy is five times larger in 2050 than in 1990. Despite the greater involvement and the social attitude of the citizens, the relative income differences between the Organisation for Economic Co-operation and Development and the rest of the world still exist. The GDP of The Netherlands is four times that of 1990. The per capita income in The Netherlands in 2050 is more than three times higher than in 1990. People go on holiday several times a year. Holiday destinations close to home are particularly popular, but long-distance travel is still in demand. Passenger air transport has increased by a factor of 5 compared with 1990.

The use of cars has risen by 40 per cent per person. Due to the large increase in the population, the total number of kilometres driven in cars is 80 per cent higher. In relation to 1990, the use of public transport has increased by 40 per cent per inhabitant. In total, public transport is used 80 per cent more than in 1990. The bicycle is used for shorter distances. The majority of passenger and freight transport takes place by car and truck, but nine times as much is transported by rail and one and half times as much by inland shipping compared with 1990. Agriculture is considerably less intensive than in 1990. Some 10 per cent less space is used for agricultural purposes in 2050 than in 1990 and, particularly in the cattle-breeding industry, production is lower. Mainly ecological products are consumed and the consumption of meat is lower than in 1990.

Energy and environment

Due to the intensive focus on nature and the environment, environmental problems are seen in the broader context of sustainable development and society. Solutions are first sought at a local or regional level and are aimed at combining the gradual adaptation of behaviour and technology improvements. A great deal of energy is devoted to securing the necessary broad public support.

The world's energy systems differ according to the availability of natural resources in the region. In The Netherlands region, the development of cleaner and more efficient technologies is driven by the notion that we must be economical with energy and material supplies. As a result of the moderate technological advances in energy recovery, the energy prices are relatively high. This is also a motivation for energy saving.

In order to reduce CO_2 emissions, people are prepared to accept measures that can significantly affect their everyday behaviour. People are willing to use public transport and there is a general willingness to pay higher costs for small-scale sustainable energy supplies and ecological products. At the same time, there is rapid diffusion of energy technology with lower emissions (fuel cells, clean energy carriers, combined heat and power), so that local pollution does not occur as readily.

Space

In this vision, people live in busy, compact cities. Within the cities, the concentrated built-up areas are linked up to public transport. Living and working are highly integrated in this vision. The car retains its popularity, but public transport plays an important role in passenger transport. Both the number of motorways and the number of railways have increased since 1990. Air traffic has lost ground to the train for shorter journeys. In this vision, The Netherlands has two national airports to meet the demand for air travel. Due to the large service sector, there are many more offices than in 1990. The space for living and working is used more efficiently but the large population lays claim to the available space. The area devoted to agriculture in 2050 is more than 10 per cent smaller than in 1990. In addition to production, agriculture has an important ecological and recreational function. New features of the landscape are large fields with crops that are used for biomass production (like poplar, willow and miscanthus). Despite the growth of the built-up environment, there are more large nature areas than in 1990.

The visions compared

In vision A of the world, the emphasis is on international co-operation, whereas in B the emphasis is on regional identity and self-sufficiency. People, ideas and capital are less mobile in vision B than in vision A. There is greater technological innovation in A than in B. In A the world economy is nine times larger than in 1990, compared to a five-fold increase in B. The world's population is somewhat smaller in A than in B, although this change is far more pronounced world-wide than in The Netherlands, which has almost 3 million inhabitants more in vision B than in vision A.

Vision A portrays an individualistic, competitive society where people put their own interests above those of others. B, on the other hand, is a sociable and family-oriented vision, where one's fellow man and public interests are central. In the individualistic vision of A, the thinning out of the family will take place far more quickly than in the socially-minded vision B. The ageing of the population also takes hold quicker in A.

Nature and the environment are used as a springboard for progress in A, and it is mainly for this reason that they are important. In B, nature and the environment come first. The aim is a sustainable society, which does not impose a strain on the environment. Market mechanisms dominate in A, while in B there is strict regulation and distribution of wealth; a consensus dialogue model is favoured for dealing with various socio-economic issues. Vision A, on the other hand, gives few opportunities for participation and has a very hands-off government.

The Dutch economy in A is more than five times larger than in 1990, compared to growth of more than a factor of four in B. However, the population in B

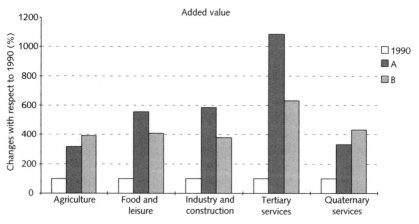

Figure 2.1 *Breakdown of projected economic growth for several main sectors in The Netherlands in the two visions.*
Economic growth (value added) is considerable in all sectors, but particularly strong in the (energy- and material-extensive) service sectors. Growth in industry is to a large extent the result of a shift from base materials to products with a high added value. The growth in physical production (tonnes of products) is therefore considerably lower than the rate of economic growth.

is 15 per cent larger than in A. The people in A are thus one and a half times as rich. Due to the larger population size in B, absolute consumption is only a quarter lower than in A.

In B, recreation is largely enjoyed close to home and holidays in The Netherlands are also popular. In A, international travel is part of everyday life throughout the whole of society. In A, people fly two and a half times as often as in B. In B, public transport is available on a much wider scale than in A, and an individual in B travels at least one and a half times more often by public transport. In A, people use their cars some 15 per cent more often than in B, but because there are around 15 per cent more people in vision B they drive just as many kilometres as in vision A. Rail transport and inland shipping are growing faster in B than in A. Due to the importance attached to nature, cattle breeding in vision B is animal-friendly and extensive. In vision A, agriculture is highly intensive and the well-being of the livestock is not of prime importance.

In B, people are prepared to spend more and to change their behaviour to arrive at a sustainable solution. People in A act as 'homo economicus' and are not easily prepared to change their behaviour on the grounds of principle. Measures must have as little effect as possible on their prosperity and consumption patterns.

The people in A live in large houses in spacious suburbs, while in B the people live in compact cities and in smaller houses. There are many motorways in A, while in B there is more of a railway infrastructure. In addition to production, the extensive agriculture in B also has an ecological and recreational function. In B there are large nature areas. The nature areas in A are fragmented by roads, agriculture and built-up areas.

Two energy systems to realize an 80 per cent reduction of greenhouse gas emissions

The energy system of vision A

World trade increases drastically in this vision. The dynamic economic development ensures that there is rapid technological development in many areas. People prefer luxury and comfort and earn more than enough to pay for it. The high incomes of people and the importance of material wealth make it unlikely that lifestyles will be adapted solely in response to environmental arguments. The design of the energy system is largely determined by cost optimization without taking the environmental costs into account. Relatively expensive (energy) options (both for enhancing energy efficiency and energy supply options) are therefore not favoured.

There is a preference for cheap energy carriers. Because of strong competition on the world energy market there are large quantities of cheap fossil fuels available, particularly coal and natural gas (the latter partly from unconventional sources such as Coal Bed Methane and methane stored in clathrates, particularly in deeper sea and ocean waters). Coal, with cost levels at around US$1/ GJ, is a preferred fuel, and can be used for the production of both electricity and hydrogen.

The design of the energy system is largely determined by market forces and cost minimization is an important driver. This implies a tendency towards larger-scale production capacity rather than a decentralized energy (electricity) supply system. This production capacity is built close to industrial centres for optimal heat utilization.

Advanced power plants with efficiencies in the range of 50-70 per cent can easily be equipped with CO_2 recovery steps with only moderate economic and efficiency penalties compared to their use without CO_2 capture and storage. Very important is the production of hydrogen from coal to be used for transportation purposes. Carried out on a large scale and combined with CO_2 removal, the hydrogen is carbon-neutral and can be produced with high energy-conversion efficiencies of about 60 per cent.

Box 2.1

Some key technologies in vision A

Transport

Energy for transport has come to form a major part of the energy demand. Fuelled with hydrogen, fuel cell vehicles (FCVs) are more efficient than the internal combustion engine vehicles that dominate the current transportation fleet. However, the production of hydrogen requires more primary energy than diesel or petrol. Nevertheless, the overall energy (and CO_2) balance is positive. Another very important characteristic of FCVs is their contribution to reducing other emissions such as NO_x and dust, which results in greatly improved (urban) air quality.

Energy system

Production of hydrogen and power from coal with CO_2 removal and storage; fossil fuels, in particular coal and methane derived from coal beds, still play a key role in meeting the energy demand. Sustainable and competitive use of fossil fuels is possible with advanced technologies such as gasification, high temperature gas separation and advanced power cycles. Such systems allow for cheap and efficient removal of CO_2, resulting in carbon-neutral power and hydrogen. Energy efficiencies of these systems are high (50-70 per cent) when applied on a large scale (e.g. 1000 MWth input).

Industry

Overall efficiency improvement in heavy industry combined with material savings; since in this vision industrial production grows considerably, increasing energy efficiency and (raw) material efficiency are important aspects of keeping the energy demand within limits.

Built-up environment

Energy-efficient dwellings: if no measures are taken, energy use in the built-up environment will increase drastically due to more luxurious standards of living and larger dwellings. Furthermore, these houses are built in suburban environments, making application of district heating (CHP) very difficult; increased energy efficiency of dwellings (in particular heat demand) is therefore very important. High levels of thermal insulation, heat recovery equipment and widespread application of solar heating keep the final energy demand of the built-up environment within acceptable limits. The remaining heat demand is met with efficient heat pumps.

A significant share of 'coal-based' power generation and hydrogen production comes from Coal Bed Methane. Coal Bed Methane is produced by injecting CO_2 in deep coal layers. CH_4 present in those coal layers is displaced by the CO_2, resulting in a CO_2-neutral gaseous energy carrier.

Nuclear energy also plays a role in the future energy mix, particularly for power generation (30 per cent). Waste heat from nuclear power plants is also used for industrial heat and district heating. Due to the development of inherently safe reactor designs and new methods for treating nuclear wastes (rapid breakdown of long-life isotopes), public resistance to this energy option will have diminished by the year 2050. The low cost of power generation is another argument for the introduction of nuclear reactors. Furthermore, these reactors do not depend solely on uranium for their fuel requirements, but can also be fuelled by thorium.

Electricity probably has to be imported, especially since by the year 2050 a liberalized open energy market has led to a fully functional European grid and free trade in power and gaseous energy carriers. Nuclear-based electricity from France, and possibly also surplus wind power from the UK, can contribute to meeting the Dutch demand.

Another major energy supply option is the import of bio-energy. Produced cheaply in large areas of Eastern Europe, Latin America and Sub-Saharan Africa, both raw biomass and energy carriers from biomass (hydrogen in particular) find their way into the Dutch energy system. Biomass production costs could be in the same range as coal costs. Because there is no need for CO_2 removal and subsequent storage, bio-energy is competitive with fossil fuels. Biomass is used mainly for production of synfuels, and partially for the production of electricity and some feedstock for the chemical industry. Mineral oil remains an important feedstock for the petrochemical industry but not for transportation fuels.

Renewables such as wind and PV play only a modest role. Wind energy is generated mainly offshore in large wind parks but total production makes only a relatively modest contribution of 10 per cent to the electricity supply. The application of PV is limited to niche markets since it generally supplies more expensive electricity than the large centralized capacity. Solar heating makes a very significant contribution by meeting a substantial part of the demand (about 100 PJ) for low-temperature heat in dwellings and offices. The contribution of other renewables is negligible.

The energy system in vision B

There are two very important aspects of the energy system in the B vision. First, the total energy demand in terms of primary energy carriers has not increased due to developments in the economic structure (reduced energy intensity) and far-reaching energy-efficiency measures throughout the economy. The more modest growth in energy consumption and growth of economic activity compared to vision A puts less strain on the energy infrastructure (distribution of power, heat, gas and fuels). Consequently, the capacity of the energy infrastructure in place at the beginning of the 21st century remains sufficient. This results in a preference for adapting the existing energy infrastructure rather than designing and installing a completely new system.

Secondly, in principle, society prefers renewable energy sources (wind, solar energy and bio-energy). Since renewables (solar and wind in particular) play an important role in the energy supply, and specifically in power generation, the energy system must be able to cope with the intermittent character of those supply options. This involves installing storage capacity and more especially investing in flexible power-generation capacity fired with natural gas (without CO_2 storage, since this would be uneconomical for gas-fired capacity with low load factors).

An advantage of the large share of renewables (linked to a European grid) is that the fluctuations are less severe than for smaller shares linked to a local grid.

The main characteristic of this vision is the strong support for renewables. Nuclear energy is largely missing from the (Dutch) energy-supply mix and the use of fossil fuels is kept to a minimum.

This leads to an impressive increase in the wind-power capacity, both on-shore and especially offshore. The longer term potential of wind-power generation is about 15,000MWe installed capacity. This represents a net contribution to the power generation of about 40 per cent, making wind the single most important source of energy for electricity generation. Another highlight of the power generation in this vision is the structural, large-scale application of solar energy, which is largely concentrated and integrated in the built-up environment (in industry, urban areas and some applications in agriculture). Photovoltaic energy is fully integrated in the built-up environment. PV is still a relatively expensive option for power generation in 2050, but the higher costs are accepted by society. They are partially incorporated in the prices of houses and offices. The contribution of PV to the total power requirement is around 15 per cent.

Solar heating contributes substantially to meeting demand for low-temperature heat (around 150 PJ primary energy).

The bulk of the remaining energy demand is met by bio-energy, which meets two-thirds of the national energy demand. 'Bio-energy refineries', where bio-alcohols are produced as well as hydrocarbons to partially cover the feedstock needs of the chemical industry (the remaining demand is still covered by mineral oil), are the main consumers of biomass. Some of the energy output from these integrated complexes is heat and power. Such complexes are typically built in large, built-up sea port areas where biomass can easily be supplied from all over the world. This reduces supply risks caused by fluctuations in biomass production in different regions. Part of the biomass-fired capacity can also be fed with fossil fuels when needed.

A small part of the biomass (about 15 per cent of the total primary energy demand) originates from indigenous sources, such as (the organic fraction of) waste and agricultural residues. More of it is cultivated, particularly in areas where the growing of biomass can be combined with other land-use functions, such as buffer zones for nature areas and sound insulation barriers. In the case of forestry, the supply of timber, pulpwood and biomass fuel is combined. In total, about 200 PJ can be produced from indigenous biomass.

However, the bulk of the biomass is imported. Imported bio-energy includes raw biomass (for large-scale biomass conversion capacity in sea ports) and biofuels. Biomass is supplied from various regions around the world, with Eastern Europe, Latin America, Canada and sub-Saharan Africa being the most important

Box 2.2
Some key technologies in vision B

Transport

Energy use for transport has come to form a major part of the energy demand. Through catalytic oxidation or reforming, hydrocarbons and alcohols can be used for firing fuel-cell vehicles. Fuel-cell vehicles (FCVs) are more efficient than internal combustion engine vehicles. However, including the production energy of the fuel used, the overall energy balance for some vehicles (e.g. the truck) is negative. Nevertheless, for alcohols derived from renewable energy sources, the CO_2 balance is positive. Another very important characteristic of the FCVs is their potential contribution to reducing other emissions such as NO_x, dust and N_2O, which results in highly improved (urban) air quality.

Bio-energy

Production of transportation fuels and feedstock from biomass; the societal preference for renewable energy sources in this vision makes biomass the most important primary energy source. So-called 'bio-refineries' are capable of converting various biomass feedstocks (from clean wood up to mixed waste) to bio-crudes, which are suitable as feedstock for the chemical industry, and bio-alcohols. Large amounts of bio-alcohols, in particular ethanol and methanol, are used for transport. Apart from fuels and feedstocks, bio-refineries also produce power. Both biochemical processes (for production of ethanol) and thermochemical processes (syngas production and conversion to methanol and Fischer Tropsch liquids) are applied. The process integration ensures a high overall energy efficiency of such factories. Located in sea ports, those facilities can be fed with imported biomass. Besides raw biomass, biofuels which are produced in other areas of the world are also imported.

Electricity production

Electricity production from wind and solar energy; the contribution of solar energy and wind energy to electricity production is maximized. 15,000 MWe of wind power, largely placed in offshore wind parks, and structural application of photovoltaic energy in the built-up environment, make a considerable contribution to the primary energy supply. Due to the intermittent character of solar and wind, some storage capacity (batteries at local level for PV, compressed air facilities in combination with wind park) and peak power generation capacity (natural gas-fired gas turbines and fuel cells) are required to guarantee a stable electricity supply. But integration in the European grid and the large-scale application of those renewables keep the need for peak power capacity within reasonable limits.

Built-up environment

Energy-efficient dwellings: Compact cities and buildings, high insulation grades, heat recovery equipment and widespread application of solar heating keep the final energy demand of the built-up environment (both dwellings and offices) at very low levels. Passive solar heating and solar collectors cover most of the remaining heat demand.

suppliers. The Netherlands requires roughly four million hectares abroad to produce the biomass and bio-fuels needed to meet the national energy demand.

Natural gas still plays an important, although relatively modest, role in the
energy mix: flexible peak power-generation capacity (highly efficient gas turbines
and fuel cells) covers the remaining part of the power production. This capacity
makes the large contribution of the intermittent wind and solar-driven power-generation capacity feasible. CO_2 removal and subsequent storage is not considered
for this capacity due to the relatively small capacity and low load factors.

Finally, part of the gas supply is covered by Coal Bed Methane, which is produced by injecting CO_2 in deep coal layers. Overall, the use of methane is carbon-
neutral since roughly two units of CO_2 are needed to produce one unit of methane.
The production potential of Coal Bed Methane in The Netherlands is supposed to
be considerable.

In order to operate such a system some CO_2 removal capacity is needed for
large-scale energy conversion systems. Part of the CO_2 is recovered at ammonia
production plants and other industries where hydrogen is needed and CO_2 is a by-
product. Some CO_2 is collected from the bio-refineries. As a consequence, CO_2
storage is less important than in vision A. It is only applied in industrial centres
(such as the Rijnmond area) and at baseload power generation capacity.

A comparison of the energy systems in the two visions

This section summarizes some of the main choices, differences and quantitative
aspects to be found in the envisioned energy systems of the future. Figure 2.2
summarizes the extent to which greenhouse gas emissions would be reduced with
the various main categories of options (efficiency measures, renewables, CO_2 storage) compared to 1990 levels and a baseline development where the growing energy demand arising from economic growth is covered by the present-day fuel mix
which is dominated by fossil fuels. Clearly, greenhouse gas emission levels will
soar to unacceptable levels if nothing is done to counteract increasing fossil fuel
use to drive economic growth. On the other hand, the two very different visions and
the strategies that it is assumed will be implemented do meet the target of 80 per
cent emission reduction.

Table 2.2 shows that improvements in energy efficiency in all sectors of society are a key factor in achieving this. Even though economic growth is very high over
the time-frame considered, in both visions the total primary energy consumption
is reduced in relation to 1990. The reduction is particularly extreme in the built-up
environment, where energy savings on low-temperature heat for space heating and
warm water as well as highly efficient electric appliances reduce the net energy use
by around 80 per cent. Only in the transport sector does energy use rise in absolute
terms compared to the reference year, despite the introduction of much more efficient vehicles and transport systems, as such improvements cannot offset the projected overall growth in transport, particularly freight transport and air traffic.

Except for transport, all sectors are able to reduce their total energy demand
due to strong energy-efficiency improvements. For transport, energy efficiency
improvement is unable to compensate for the (strong) growth in total demand for
transport services.

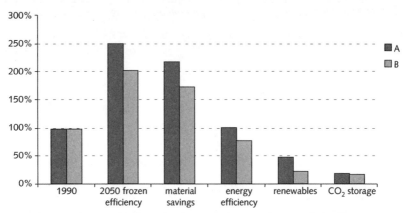

Figure 2.2 *The contribution of different kinds of measures to reduction of greenhouse gas emissions in relation to the 'business-as-usual scenario' for both visions (100 per cent means the total greenhouse gas emissions (in CO_2-equivalents) in 1990). The emission levels shown under 'frozen efficiency' assume no improvements in energy efficiency and the same fuel mix to meet energy demand as in 1990.*

Table 2.3 illustrates how the primary energy demands in each of the visions compare with each other and with the 1990 energy consumption. A combination of structural changes in the economy (more emphasis on services and light industrial sectors) and dramatic improvements in energy efficiency make it possible that the total energy demand decreases.

Obviously, the energy mix in both visions has changed dramatically compared to the current situation. The contribution of renewables, in particular (imported) bio-energy, has increased by a factor of 25 (vision A) or 40 (vision B). Consequently, the share of fossil fuels has declined, in particular mineral oil. The use of coal (although this could also be true for natural gas) differs greatly between the two visions, although this fuel is always assumed to be used in combination with CO_2 removal and storage.

Table 2.4 summarizes the key characteristics of the two visions, with an emphasis on the assumptions made and the technologies related to the energy systems that are assumed to be in place. This summary illustrates that the goal of reducing emissions by 80 per cent can be achieved with quite contrasting strate-

Table 2.2 *Main demand sectors and their share in the total primary energy demand for The Netherlands in 1998 and in the two visions.*

Demand sector	Netherlands 1998 (PJ)	Vision A (PJ)	Vision B (PJ)
Industry	1203	685	503
Transport	451	874	975
Domestic	538	164	105
Services	482	250	119
Agriculture	192	22	66
Total energy use (approx.)	**3000**	**2000**	**1770**

Table 2.3 *Simplified energy supply mix for The Netherlands in 1998 and the two visions*

Fossil fuels	Netherlands 1998 (PJ)	Vision A (PJ)	Vision B (PJ)
Natural gas	1480	170	130
Mineral oil	1030	220	140
Coal	380	580	150
Subtotal fossil fuels	2890	970	420
Nuclear energy	25	80	-
Renewable energy			
Biomass & organic wastes	30	800	1180
Wind energy	5	30	80
Solar energy	-	110	90
Subtotal renewables	35	940	1350
Total energy use (approx.)	**3000**	**2000**	**1770**

gies. Certain aspects appear in both visions (such as increased energy efficiency and mitigation of non-CO_2 greenhouse gases). Other aspects may be competing or even contradictory (such as the energy infrastructure required) and raise the question of how such developments can be steered over time and whether it is possible to determine the optimal strategy in advance.

Table 2.4 *Summary of the key characteristics of the assumed energy systems of the two visions*

A	B
Energy demand	
Strong growth of industrial production capacity, luxury dwellings in suburban areas and strong growth of individual transport and air traffic are dominant developments affecting the energy demand. With a total population of 16 million people, the total energy demand amounts to 2000 PJ; industry and transport are the dominant sectors in energy use.	More limited growth of industrial production, considerably increased material efficiency, compact cities with highly energy-efficient dwellings, more intensive use of public transport limit the energy use. On the other hand, there are a larger number of people living in The Netherlands, 19 million compared to 16 million in A. Total energy demand amounts to over 1700 PJ; energy use is dominated by the transport sector and to a lesser extent by industry.
Built-up environment	
Large, luxury dwellings in suburban areas. Strong growth of the service sector. Considerable improvements in energy efficiency through good insulation. Energy demand covered principally by electricity and heat pumps and passive solar energy and solar heating systems.	Compact cities and very energy-efficient, compact dwellings. The service sector is dominant sector in the economy. Offices are highly energy-efficient. Energy demand covered principally by heat distribution, solar heating and large-scale application of PV in the built-up environment.

Table 2.4 continued

A	B
Industry	
Substantial improvements in energy efficiency by new industrial processes. Largely driven by the need to reduce production costs and improve product quality. Industrial capacity is organized in integrated industrial complexes with very efficient combined heat and power generation and optimal waste heat utilization. Material efficiency is also improved.	Strong improvements in energy efficiency obtained with far-reaching measures, including expensive options to reduce energy consumption. Well-developed recycling schemes and adapted products lead to substantially increased material efficiency (up to 40 per cent compared to baseline levels). Industrial production capacity is concentrated in integrated complexes with highly efficient CHP generation and heat cascading.
Transport	
Dramatic increase of air traffic and freight traffic in general. Strong growth of individual transport. Minor role of public transport. Cars and trucks powered by very efficient and clean fuel cell vehicles driven by hydrogen. Efficient aircraft are fuelled by synfuels produced from CO_2-neutral biomass.	Increase in the need for transport in general, but major role for public transport in passenger transport as well as train use for freight. Cars and trucks powered by efficient and very clean fuel cell vehicles fuelled by bio-alcohols.
Agriculture and land use	
Highly intensive agricultural production over a limited surface area. Very efficient greenhouse sector and intensive cattle raising in 'closed stables'. Considerably increased land use for suburban developments (comfortable dwellings in park-like environment). Increased land use for industry and infrastructure (roads, airports). Other land areas important for recreational purposes.	Ecological and more extensive agricultural practice. Integration of functions: agriculture, recreation as well as energy production; land is also used for biomass production for energy and feedstock. Compact cities and large-scale application of public transport.
Non-CO_2 Greenhouse Gases	
Far-reaching reduction of HFCs through application of alternative compounds, elimination of N_2O emission of combustion-related emissions through application of Fuel Cell Vehicles and other industrial processes. Strong reductions of N_2O and CH_4 emissions in agriculture through closed stables and greenhouses.	Far-reaching reductions of HFCs through application of alternatives, elimination of N_2O emission of combustion-related emissions through application of Fuel Cell Vehicles. Strong reductions of N_2O emissions through less intensive agriculture. Some remaining CH_4 emissions related to cattle raising.

Table 2.4 continued

A	B
Energy supply	
Energy supply dominated by coal and biomass. Furthermore, a diverse mix including nuclear, natural gas and Coal Bed Methane and wind power. Mineral oil remains of importance for the chemical industry.	Energy supply is dominated by biomass (large-scale import of raw biomass and biofuels from other parts of the world), about 15 per cent of the primary energy comes from indigenous biomass and (organic) wastes. Remainder of energy supply covered by wind, solar energy, coal (Coal Bed Methane) and natural gas.
The energy system	
A completely revised energy infrastructure with a large-scale hydrogen distribution network and a network of CO_2 pipelines for collection and storage of CO_2. CO_2 removal is especially applied at hydrogen production and power generation from coal and a number of heavy industries. The energy supply (power, fuels, heat) is dominated by large scale capacity. Energy production is to a large extent integrated in industrial complexes for optimal utilization of waste heat. The energy supply in the built-up environment is dominated by electricity, application of heat pumps and passive solar energy.	A more traditional energy infrastructure remains in use: large-scale heat distribution is used for compact cities. CHP generation and heat cascading is widely applied in integrated industrial complexes. The dominant role of wind and solar energy in the electricity supply requires storage capacity (batteries in the built-up environment, compressed air for wind parks) and flexible, gas fired power generation capacity. PV and solar heating are widely applied in the built environment.
Electricity	
Electricity production dominated by coal with CO_2 removal and storage, nuclear energy. Remaining part by natural gas + Coal Bed Methane and wind power.	Electricity production for more than half covered by wind and solar (PV) energy. Supportive peak power generation capacity on basis of natural gas and CO_2-neutral Coal Bed Methane.
Fuels	
Large-scale production of hydrogen from coal as well as from biomass. Biomass is also the CO_2-neutral feedstock for production of synfuels, particularly for use in air transport. Substantial import of bio-energy.	Vehicles are fired by bio-alcohols produced from indigenous biomass and large-scale import of biomass and fuels from various world regions. Biomass is also used for production of synfuels, for use in the air transport sector and as feedstock in chemical industry.

Concluding remarks: implications for changes

The exercise above shows a number of things very clearly. In the first place, without measures to limit energy use and promote the use of energy sources with at best very low CO_2 emissions and with steady economic growth over a long period of time, greenhouse gas emissions will soar to unacceptable levels. Secondly, the two fairly contradictory visions show that it is feasible to combine (strong) economic growth with far-reaching reduction of greenhouse gas emissions provided a number of key options are developed and applied. Thirdly, the results also show that such a result can be achieved in different ways. In other words, different, and in a number of respects opposing, strategies are feasible for arriving at a low greenhouse gas emission economy in the long term.

Clearly, the number of (often bold) assumptions made in the visions illustrate that drastic changes in many areas of society are required to reach the greenhouse gas emission levels desired in 2050. However, the changes required may differ considerably depending on which vision one considers most feasible and the context one assumes for the longer term. And changes are extremely complex. Questions immediately arise about how major transitions are to be realized. Different technologies and complete energy systems and infrastructures compete. In addition, technologies become available over different time spans, greatly depending on R&D efforts and (policy) support or international developments. Who will pay the inevitable (additional) costs of transforming the energy system? How are the potential changes to be initiated, implemented and steered?

The visions could be elaborated in far more detail but that is not the objective of this book. The visions serve as illustrative and provocative examples for the later chapters. The visions are the basis for reflecting on the transformations and reversals of (current) trends that are needed to meet the ultimate objective. The next chapter presents an overview that gives some insight into difficulties in realizing crucial changes. Chapters 4 to 11 of this book will explore and discuss the selected issues, fields and sectors in detail. They cover social sectors such as households and agriculture, specific options for reducing emissions (illustrated by improving material efficiency as a means to reduce greenhouse gas emissions) and policy areas such as local and international climate policy, technology management and R&D policy and the effectiveness of various policy instruments. The discussion evaluates the feasibility of realizing the drastic changes required to create a climate-friendly society, as well as the prospects of and barriers to success, from very different perspectives. The combination of the findings can provide an insight into transition processes and the full extent of their complexity. An attempt at such a synthesis will be made in the final chapter of this book.

References

Alcamo, J., A. Bouwman, J. Edmonds, A. Grubler, T. Morita and A. Sugandhy, An evaluation of the IPCC IS92 Emission Scenarios. In: Houghton, J.T., L.G. Meira Filho, J. Bruce, Hoesung Lee, B.A. Callander, E. Haites, N. Harris and K. Maskell, eds., *Climate Change 1994: Radiative Forcing of Climate Change and An Evaluation of the IPCC IS92 Emission Scenarios*, Cambridge: Cambridge University Press for the Intergovernmental Panel on Climate Change, 1995.

Beer, de, J.G., *Potential for Industrial Energy Efficiency Improvement in the Long Term*, Ph.D.-thesis, Department of Science, Technology and Society, Utrecht University, 1998.

Beer, J.G., M.T. van Wees, E. Worrell, K. Blok, *ICARUS 3: the potential of energy efficiency improvement in The Netherlands up to 2000 and 2015*, Department of Science, Technology and Society, Utrecht University.

Faaij, A., S. Bos, J. Oude Lohuis, D. Treffers, C. Battjes, J. Spakman, R. Folkert, C. Hendriks, *Sustainable energy systems on the long term - two visions on the Dutch energy system*, report prepared for the COOL dialogue by the Department of Science, Technology and Society, Utrecht University, Netherlands Energy Research Foundation, State Institute for Health and Environment and Ecofys. Project supported by the National Research Programme on Global Air Pollution and Climate Change, 1999.

Faaij, A., R. Van den Broek, B. Van Engelenburg, E. Lysen, *Global availability of biomass for energy and possibilities and constraints for large scale international trade*, The Fifth International Conference on Greenhouse Gas Control Technologies, 13-16 August 2000, Cairns, Australia.

Farla, J., *Physical Indicators of Energy Efficiency*, Ph.D.-thesis, Department of Science, Technology and Society, Utrecht University, 2000.

Grübler, A., et al., *Global Energy Perspectives to 2050 and Beyond*, World Energy Council (WEC)/International Institute for Applied Systems Analysis (IIASA), 1995.

IPCC, *Special Report on Emission Scenarios*, Cambridge: Cambridge University Press, 2000.

Nakicenovic, N. et al., Energy Scenarios, in: *World Energy Assessment of the United Nations*, UNDP, UNDESA/WEC, Published by: UNDP, New York, 2000.

Phylipsen, D., *International Comparisons & National Commitments – Analysing energy and technology differences in the climate debate*, Ph.D.-thesis, Department of Science, Technology and Society, Utrecht University, April 2000.

Rogner, H. et al., Energy Resources, *World Energy Assessment of the United Nations*, UNDP, UNDESA/WEC, Published by: UNDP, New York, September 2000.

Treffers, D.J., A. Faaij, E. Lysen et al., *Database Clean Energy Supply 2050 - Final Report*. Department of Science, Technology and Society, Netherlands Energy Research Foundation, ECOFYS - under coordination of the Utrecht Center for Energy Research. Carried out for the Ministry of Housing, Spatial Planning and Environment (VROM), 2001.

Turkenburg, W.C. et al., Renewable Energy Technologies, *World Energy Assessment of the United Nations*, UNDP, UNDESA/WEC, Published by: UNDP, New York, September 2000.

Chapter 3

Contemporary practices: greenhouse scepticism?

Walter Vermeulen and Marcel Kok

In this chapter...

... having looked ahead to 2050 in the previous chapter, we now return to the present situation in The Netherlands. Climate change has been addressed in Dutch environmental policies for over a decade. This chapter considers the societal response to these policies. Has the increased policy attention led to changes in the patterns of behaviour of firms, authorities and consumers? We can observe a widespread recognition and acceptance that climate change is a problem that has to be dealt with. But we can also see a gap between the willingness to act at the political level and the slow change in actual behaviour of target groups and the absence of climate change as an issue in their daily lives. This forms the national context in which climate policies are developed and it poses several challenges for developing long-term climate change policy. The analysis in this chapter will result in the formulation of a number of dilemmas that need to be addressed in thinking about such long-term strategies. In subsequent chapters the authors will reflect upon these dilemmas.

Introduction

The scenarios sketched in Chapter 2 are just two contrasting examples from a wide variety of conceivable routes towards a climate-neutral society. Whichever route is taken, they all embody manifold drastic transitions within the period of 50 years; technological, institutional, structural, behavioural and cultural.

Such societal transitions will require collective decision-making in the political system, but equally decisions and changes in behaviour by many actors in the sectors of society discussed above. Everyone will have to push in the same direction and that will not be easy to achieve. We know that societal responses vary from denying or ignoring the environmental and societal effects of climate change to adapting or trying to prevent them (Rayner, 1998). Not everyone will necessarily wish to prevent the impact of climate change and to carry the (potentially heavy) burden of a forced transition. Adapting to climate change or even ignoring it may fit very easily into the political culture of substantial groups in society, labelled as fatalists and individualists (Schwarz and Thompson, 1990) or as cornucopians in their attitudes to addressing environmental or climate change issues (see Thompson and Rayner, 1998). The fatalists' and individualists' response to threats of nature like climate change is that nature is capricious or benign. The cornucopians' response is that nature is intrinsically robust and overflowing with an abundance of resources. Both responses result in sceptical attitudes.

It may very well be that the political elite of western countries agrees upon the direction and pace of reaching climate goals. However, to get there, they depend heavily on the day-to-day decisions of many different actors, among whom there still is a distinct possibility of meeting scepticism and unfamiliarity with the issue. So the central question at this stage of this book is: what do we know about societal support, about awareness and about contemporary practices in efforts to change the behaviour of individuals and organizations in response to climate change? And if we are confronted with scepticism and unfamiliarity, can we deal with it? Or should we and can we design routes of transition independently of these groups' co-operation? What would that imply for policy scenarios?

In this third chapter we will evaluate the current societal support for climate change policies and the contemporary practices of various actors within society using a large number of studies and evaluations available for The Netherlands. We will examine whether, and if so how, climate change issues play a role in decision-making amongst societal actors in relation to their other interests. We will show that there is a gap between the willingness to set (modest) targets at the political level (as expressed in various national climate change policy documents) and the slow pace of change in the actual behaviour of target groups (both government agencies and in society). In the last section, we will conclude this chapter by identifying seven dilemmas that are relevant in exploring possible transformation processes and which will be taken into account in the following chapters.

Actors: who decides on innovations?

As Chapter 2 illustrated, a shift towards a climate-neutral society will require the efforts of industries, companies in the energy production and distribution sectors and agriculture. Substantial efforts will also be needed in the fields of transportation and household consumption. Making the transition to a climate-neutral society will not only affect the daily routines and technologies adopted in specific economic sectors, but also represent a challenge for complex processes of collective decision-making, for example in the areas of building and infrastructure projects.

Some of these transitions are large-scale innovations involving few decision makers. In such cases, businesses depend on co-operation with the national government for approval of projects and perhaps financial support (e.g. CO_2-recovery/storage or large-scale wind parks at sea). Decision-making may then be relatively well-ordered and rest mainly on techno-economic assessments, although local opposition may be encountered and public support might become a critical issue.

But in most cases the transitions depend on far less structured decision-making mechanisms, either involving individual firms or households that do not actually yet address the issues concerned (e.g. applying eco-design principles or using solar energy appliances) and for whom changing the relevant routines is more than just a matter of making a techno-economic assessment. As Table 3.1 shows, most of the transitions do not involve large-scale projects involving a few actors from business and central government, but fall into the category in the bottom right corner.

Table 3.1 *Scale of the innovative product and number of decision makers involved in transitions*

Scale of product	number of decision makers		
	few, independent	*few, mutually dependent*	*many, independent*
Large scale	▪ application of wind (at sea) ▪ CO_2-recovery/storage ▪ bio-refinery	▪ infrastructure projects	
	▪ bio-industry ▪ 50% more eco-efficient aviation	▪ application of wind ▪ sustainable technological development policy	▪ application of biomass ▪ application of CHP ▪ hydrogen society
small scale		▪ fuel cells in motor vehicles ▪ widespread re-use/ recycling ▪ low impact construction ▪ supply of eco-designed appliances	▪ improving material efficiency ▪ improving energy efficiency ▪ application of sun ▪ eco-agriculture ▪ consumers' choices

This implies that achieving the goal of reducing emissions by 80 per cent mainly depends on the acceptance and willingness of many businesses, local institutions and individuals to contribute to the transitions. This in turn will depend on the perceptions these various groups in society have of their own contribution to the problem and of their own and others' responsibility and possibilities to address it. A complicating factor here is that the suggested options are mutually dependent. For example, the degree to which producers will apply eco-design principles will strongly depend on their perception of the responses of consumers to an increased supply of low-impact appliances and consumer goods. As another example, producers' choices in the quality of their energy supply (using renewable energy or not) will have an effect on the environmental profile of their products. This may lead to a growing demand for renewable energy, which will be a strong stimulus for the energy sector to invest in renewable energy. Manufacturers can decide either to leave the choice of modes of renewable energy to energy companies or to be selective at this point and to opt for green energy. In the liberalized energy market, as consumers, businesses may even decide on the modes of energy production, e.g. by producing solar or wind energy themselves. Such *inter*dependencies and growing *in*dependencies may form either a barrier to transitions ('it is their responsibility') or a chance to accelerate the process of transition.

Having identified groups of actors that are likely to contribute to trend breaks, we will now roughly sketch the way these target groups are nowadays addressed by policies.

Contemporary policies

The societal sectors mentioned have already been exposed for some time to environmental and energy policies; policies that have been successful in some fields but not in others. We will show that changes made by these target groups are only partly induced by regulatory pressure. With respect to activities and behaviour related to climate change, the day-to-day decisions of these sectors are often not addressed by compulsory regulations. Contemporary changes in policy and behaviour of these societal sectors are often induced by coercion, voluntary agreements, societal pressure, financial stimuli and market stimuli. In this section we will first discuss the policy strategies applied in the field of energy and environment that address industry, energy production, consumers and mobility. We will then go on to discuss developments in national and local climate change policies.

Environmental and energy policy

Energy policy targeted at *industry* can focus on the energy efficiency of production processes and distribution, possible co-operation in CHP and energy supply and the choice of renewable energy. The actual policy concentrates on improving energy efficiency through voluntary agreements mainly based on energy conservation methods. These agreements are intended to be incorporated in firms' environmental permits. Research has shown that these voluntary agreements are generally effective, although the level of annual efficiency improvement (approx. 1.7 per cent in the period 1989-1995) will not be sufficient in the context of reducing emissions by 80 per cent (Glasbergen et al., 1997, p3, p76). The most recent generation of agreements is intended to have a wider scope, including wider implementation of potential methods of improving energy efficiency: life cycle management, regional co-operation (industrial ecology), reduction of use of virgin resources and increased use of renewable energy. These methods are also stimulated in *industry-targeted* environmental policy, which also involves voluntary agreements, increased flexibility of permits, application and certification of environmental management systems, stimulation of eco-design, life cycle co-operation and promotion of products with low environmental impacts.

In the field of *energy production and distribution* the institutional changes resulting from the current energy policy with its dual targets (sustainability *and* creation of free energy markets) are clearly visible. The creation of free energy markets for electricity and natural gas requires a substantial re-institutionalization, which is just now taking place. The process involves privatization, resulting in a decline in the influence of local politics. It also implies scale enlargement and internationalization of energy companies, which will possibly affect the environmental policies and practices that these companies developed during the 1990s. On the other hand, new actors are entering the market place, such as retailers and housing agencies selling energy, sometimes solely green energy.

Current policy targets are to increase energy efficiency by 33 per cent in 25 years, the use of renewable energy for 10 per cent of energy demand in 2020 and application of clean technology, combined heat and power (CHP) and CO_2 storage when coal is used for electricity production. In 2001 emissions of CO_2 are still increasing despite the success of the voluntary agreements on energy efficiency in achieving the agreed targets and the reduction in the domestic use of natural gas. Although there has been an increase in the use of renewable energy, it has not been sufficient to meet the policy objectives. Efforts to increase the application of renewable energy have met with a number of problems: local opposition (to wind parks), design and planning processes and procedures in the field of building that do not make provision for it, economic barriers (pay-back times) and lower prices for non-renewable energy due to liberalization and internationalization of energy markets. In general, the role of government is diminishing, but environmental policy still has an impact through its emission targets and the sector's involvement in the production of policy .

With regard to *consumption,* all consumer goods, transportation and domestic heating are relevant. Policy has already addressed the two latter aspects for some time. With respect to domestic heating, policy focuses on energy conservation, with the emphasis on financial and communicative strategies. The same financial and communicative strategies are employed to reduce car use. In addition, improvement in the capacity and quality of public transport should further reduce the environmental impacts of mobility. Addressing citizens as conscious consumers employing their 'purchasing power' is a more recent feature of environmental policy which is still in its infancy (Vermeulen, 2000). The second half of the 1990s saw greater use of green taxation, The Netherlands now having one of the highest levels of green taxes. Rising levels of energy taxes are in force for petrol, electricity and natural gas for heating.

Efforts to reduce the impact of *mobility* struggle with the stubbornness of this issue. Reducing the impact may involve the route of addressing motorists' behaviour (applying financial, communicative and physical strategies) or influencing choices of modes of transportation, design of infrastructure, development of clean means of transportation (through support for R&D and regulations). Although these issues are all far from new in the policy arena, on neither of these aspects can substantial improvements, let alone a reversal of trends, be reported up to now, as we are continuously facing increasing levels of mobility. One small positive effect of technological improvement can be mentioned here: the last decade shows stable levels of energy consumption per kilometre by cars, while cars have become larger. There is a rise in the use of public transport, but it barely exceeds the general increase of mobility.

Even more complicated are efforts to achieve more *structural changes,* for instance in the execution of infrastructure projects or large-scale building projects. In these projects decisions are made at the national level and strong businesses are involved. But procedures are long and include many opportunities for reconsideration, both of national policies and local policy implementation, which are likely to be availed of by one or other of the many actors involved. These complex decisions usually require the management of uncertainties relating to economic, societal and ecological effects. This combination of dealing with uncertainty and repeated opportunities for reconsideration can impede the success of such complex processes.

Developments in national climate policies

In addition to general environmental and energy policies, The Netherlands has had a specific climate policy since the late 1980s. As Box 3.1 shows, we face a conflict between ambition and reality in The Netherlands, where deficiencies in implementation lead to downshifting of ambitions. During the 1970s the Dutch media regularly reported new findings about the consequences of climate change. Between 1979 and 1984 the greenhouse theory was picked up by policy makers and framed as a CO_2 problem. Between 1984 and 1988 the climate problem rose to environmental priority status, with public attention and political statements of intent both peaking between 1988 and 1990, as was also the case with respect to other environmental problems (Dinkelman, 1995; Dresen en Kwa, 2000; Van der Sluijs, 1997). The Dutch national climate policy started with the international (1988) Toronto target of reducing CO_2 emissions by 20 per cent in 2005.

After 1990 it became increasingly clear that it wouldn't be possible to realize the original targets through domestic implementation. Figure 3.1 shows the actual emissions in the period 1980-2000 as well as the expected emissions in a global co-ordination scenario up to 2030. The Dutch National Institute of Public Health and the Environment (RIVM) forecasts that with current policies the emis-

---

Box 3.1

Main elements of Dutch climate change and energy-efficiency policies

Targets: -6% CO_2-eq.in 2008-2012, compared with 1990/1995;
           (e.g. -50 Mton CO_2-equivalents compared with unchanged policies);
           10 % renewable energy in 2020, 5% in 2010;

Main elements:
- Improvement of efficiency of production and stimulation of combined heat and power (applying long-term agreements with industries, CO_2 covenant with energy producers, a benchmarking covenant and various tax measures, such as tax reduction for investments).
- Improvement of efficiency of heating (applying energy performance standards for buildings, subsidies, knowledge transfer and agreements in the field of sustainable building, exemplary and demonstration projects).
- Tax on energy (small firms and consumers).
- Improvement of energy-efficiency of vehicles (applying European covenants and regulation).
- Improvement of efficiency of domestic appliances (using subsidies and regulations).
- Stimulation of energy recovery from waste.
- Stimulation of renewable energy (research, pilot and demonstration projects, subsidies and tax measures, accelerated liberalization of market for 'green energy').
- Increase of CO_2 absorption in new forests (in The Netherlands and abroad: FACE programme).
- Reduction of energy demand of small users: households and SMEs (energy tax and information).
- Deceleration of growth of mobility and stimulation of economic driving behaviour (tax measures, road toll information).

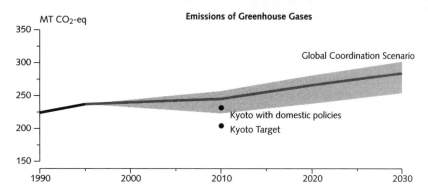

Figure 3.1 *Emissions of greenhouse gases in The Netherlands according to a Global Coordination Scenario. The shaded area depicts uncertainties in policies and societal developments (RIVM, 2000).*

sions of greenhouse gases will continue to grow. With the economic growth The Netherlands experienced over the last few years and current policies, The Netherlands can only meet its target under the Kyoto Agreement if an extra package of measures is implemented. Furthermore, reductions will have to be bought abroad (Milieuverkenning 5, 2000-2030).

The picture that emerges from this overview is, on the one hand, one of good intentions and a willingness for The Netherlands to be at the forefront of (international) climate policy-making. But on the other hand, it shows a country in a difficult position to realize its climate targets. The latter point can be explained by the economic structure of The Netherlands, with its energy-intensive industry and agriculture, the importance of transport (see also Chapter 2) and its enormous dependence on what happens internationally. In light of this situation and the awareness of the long-term necessity for far-reaching measures, in 1996 the Dutch parliament started (the first) parliamentary hearings on climate change. Parliament wanted to be in a better position to form its own opinion about the climate problem and the measures that were needed. The debate about the very existence of the climate problem and the uncertainties surrounding its causes and effects played a major role in these hearings. These hearings resulted in a political re-affirmation of the seriousness of the problem and the statement that the committee of enquiry considered a reduction of greenhouse gas emissions by 30-40 per cent in 2020 (compared to 1990 levels) to be possible.

Interestingly enough, research by Ester et al. (1999) highlights a remarkable difference between what decision-makers think the Dutch public wants and what the public say themselves about their preferences. Policy makers often complain about what they see as a gap between their habit of looking at strategic, global issues and the more locally-oriented environmental issues that occupy the minds of the general public. The argument is that citizens are only interested in environmental issues in so far as they affect their day-to-day lives. The research by Ester et al. shows that this perception is not entirely accurate. An example: all decision makers say they are well-informed about climate change issues but they believe that only a minority (35 per cent) of the public would say they are well-informed.

In fact 80 per cent say they are well-informed (whether this is correct is not impor-tant here). Even more interesting is the perceived willingness to provide extra resources for environmental protection. On this question, given a choice between seven general policy issues, 39 per cent of the policy makers put the environment in first or second place. They thought that only 7 per cent of the public would do so (and that 0 per cent would place it first). In fact, public support is far larger, as 27 per cent gave environmental protection a first or second place (Ester, 1999, p41-50).

This history shows that climate change is a widely recognized environmental issue in the national policy arena in The Netherlands. Furthermore, decision mak-ers seem to underestimate the societal support for environmental policies. But at the same time, far-reaching targets for 'a climate-neutral society' are absent from official policy documents. Such a long-term approach to the climate problem has not been widely discussed in the political debate in The Netherlands up to now, although more recently some influential advisory bodies have been demanding more attention for this aspect (VROM Raad, 1998; AER, 1999; SER, 1999). At the end of 1998 the Dutch Environmental Advisory Council published an advice enti-tled 'Transition towards a low-carbon energy system' (VROM Raad, 1998). In this advice it called for structural steps in the long-term transition towards emission reductions of 80 per cent in industrialized countries. It argued that this long-term target should drive the policy process, in which structural steps towards these goals are fundamentally more important than short-term successes. In this context, the process in the COOL project (see Chapter 11) is another interesting sign of grow-ing attention to long-term climate targets among stakeholders in The Netherlands.

Local climate policies

A similar conflict between ambition and actual practice can be observed at the local level. Local authorities can play an important role in realizing national climate pol-icy targets, but can also make more ambitious policies. As in many other countries, municipalities in The Netherlands have responsibilities and decision-making pow-ers which are essential for achieving the transitions discussed in Chapter 2. For instance, they have to issue permits for any new construction or form of land use. They are responsible for environmental permits for smaller firms. They have an important position in local traffic and public transport policies. But in all these fields their ability to pursue climate policy goals is limited because powers and responsibilities are in many cases shared with provincial and national authorities. In many cases, budgets have to be provided by national authorities. Climate change has received attention in many municipalities in recent years. Some have incorporated it in Local Agenda 21 projects, where municipalities have discussed environmental policies with local stakeholders. Other municipalities have com-mitted themselves to the International Climate Alliance of 1999, in which 109 Dutch municipalities (which is 20 per cent of all municipalities in The Netherlands) agreed to develop policies aimed at 50 per cent reduction of green-house gases in 2010 compared to 1987 levels. Members co-operate in developing possible policies and exchange information.

Although climate change discussions in Local Agenda 21 and participation in the Climate Alliance can be seen as positive signs, we must not overestimate their impact. An evaluation of the local climate policies of 51 municipalities in 1999 showed that in many cases consideration of climate change in strategic policy formulation is no guarantee of changes in greenhouse gas-related activities by local actors. There are small groups of communities actively engaged in implementing climate change strategies, but research shows that in most cases possible actions in the field of construction, traffic and green energy are not taken in practice. In many instances, basic information on sources of greenhouse gases is lacking, developments are not monitored and plans are not even developed (Burger, 1999). Frequently, a few attractive examples of activities, such as subsidizing solar hot water systems, are highlighted but there is rarely a comprehensive strategy. Research into planning procedures for large building locations and urban reconstruction projects reveals that, despite existing local policy plans, environmental and energy policies are not integrated in development procedures from the start of such projects. Many of the existing 'demonstration examples' were promoted at a later stage by actors (like energy agencies, energy companies or consultancies) who were initially not included in the process of plan development (Van der Waals et al., 1999; Van der Waals et al., 2000).

The main barrier at this stage of development of local climate change policies is (what is called) external integration of climate change issues in other fields of local policy, like traffic, economic development, urban and land use planning, housing, tax policy, etcetera. On the local level we see a considerable degree of political support for climate policies, but up to now this has only partially resulted in changes in local sectoral policies and physical changes. Opportunities for local climate policies will be discussed in more detail in Chapter 6.

Contemporary practices

There is some research into the attitudes to and public support for climate policy and contemporary practices of various groups in society. However, the picture is not complete and many different approaches have been adopted. Still, we believe that the overall picture shows a significant difference between declared public support for climate policy and daily practices and routines. We will discuss the available information in the context of public opinion, producers and consumers.

Public opinion

In the late 1990s it was often stated in public debate in The Netherlands that public concern with environmental issues was declining because of the rising prosperity. Empirical evidence doesn't support this statement when it is expressed this simplistically. Annual environmental behaviour monitoring shows that in the second half of the 1990s the Dutch population increasingly saw environmental problems as an important issue. Between 1995 and 1999 the percentage that stated that environment is 'one of the most important issues' rose from 40 per cent to 47 per cent. The number of people saying it is 'the most important issue' ranged between

17 and 20 per cent. Equally remarkable was the (small) growth in the number of people who said that environmental issues would be the greatest threat in the future: 27 per cent in 1995 and 33 per cent in 1999 (Couvert and Reuling, 2000, p24, p31, p41).

What we are actually seeing is a strong shift in the perception of environmental issues. They are decreasingly seen as threat to the health of this and future generations and increasingly people feel that 'the fuss about the environment is exaggerated' (Couvert and Reuling, 2000, p57). So public opinion is tending to shift away from an apocalyptic interpretation. We are also seeing a shift in opinions on responsibility. Dutch environmental policy is built on the principle of shared responsibility of producers, retailers and consumers. Public opinion, however, increasingly holds industry responsible and individuals are less inclined to see themselves as being responsible (Couvert and Reuling, 2000, p50-51).

This perception of personal responsibility may be the most important explanation for the gap between attitude and actual behaviour. Analyses show that this gap is smallest with respect to waste treatment and energy use and water use. The gap is widest in the spheres of food, mobility and purchasing behaviour (Hoefnagel et al., 1996; Becker et al., 1996; Steg, 1999; RIVM, 1998 a and b; Couvert and Reuling, 1999), so actual consumer behaviour is still unaffected.

At this point we should be aware of a fairly strong societal differentiation along the lines of income and even more of education. Research into environmental attitudes and behaviour shows a clear differentiation, especially according to educational levels (Hoefnagel, 1996).

The discussion above is not specific to climate change issues. There is little research specifically on the issue. One example is an analysis of public knowledge about causes, effects and solutions for environmental issues, which showed that in 1998 73 per cent were able to give a cause of climate change correctly, 65 per cent could mention a correct effect, but only 31 per cent could give a correct solution. However, looking at the full chain of causes, effects and solutions, only 25 per cent could give a proper combination, whereas 18 per cent couldn't give a proper presentation of any part of the chain (Couvert and Reuling, 1999, p39-45). This analysis covered a period of four years. It showed no significant increase in the level of knowledge during that period (Couvert and Reuling, 2000, p91). Yet other research shows that the public itself believes that of all environmental issues it is best informed about climate change, with 80 per cent saying they are well-informed or very well-informed (Ester et al., 1999, p42, p69).

With specific reference to energy conservation, research shows that the majority feel this is an important issue (56 per cent), with only 1 per cent saying it is 'not' or 'not at all' important. The number of people who feel it to be important is declining. On this aspect we also see the societal differentiation along the lines of income, and even more the level of education: people with a lower income and level of education explain energy conservation behaviour with economic arguments (squandering of energy costs money), whereas people with a higher income and level of education tend to stress environmental arguments (Couvert and Reuling, 2000, p152-163). Another study analysed priorities given to various environmental issues and showed that climate change is given low priority in public opinion: 8 per cent and 11 per cent giving it first and second place respectively.

Overpopulation (28 and 12 per cent) and air pollution (38 and 27 per cent) were given high priority by a far larger proportion of the Dutch population! Accordingly, climate change is not broadly perceived as being dangerous for human health (Ester et al., 1999, p43-44).

Why is information about public opinion important in this context? Because it is an essential resource for the necessary political process and is crucial for changing consumption patterns. Here we can see two alternative routes. The first is the route of 'conscious consumption', in which citizens may affect producers' willingness to improve the eco-efficiency of their products by purchasing alternative products with substantially lower environmental impacts. Following this route, the need for consumption is not addressed. In the second route the aim is to reduce the volume of material consumption by advocating less materialist forms of consumption or retrenching.

The latter route has been explored by Aarts, who studied rare examples of a culture of retrenching in material consumption and followed other sociological examples of refraining from consumption (like smoking). One might also expect the trickle-down effects identified by Elias to occur in the field of environmentally-relevant behaviour. Material consumption trickles down from groups with a higher social status and better education to the rest of society (Aarts, 1999, p40-41, p59-60). Her argument is that such a trickle-down effect is not likely to occur with respect to a culture of retrenchment because it lacks essential features, like the ability to distinguish yourself from others through this type of behaviour (lack of prestige) and because such an environmental attitude does not represent a cultural capacity contributing to obtaining power (Aarts, 1999; p202).

This discussion raises the issue of affecting lifestyles. Aarts cites Breemhaar et al. (1995), who showed that consumers assess their alternatives in the light of their sets of final values (like freedom, autonomy, health, personal development, social appreciation, etc.). Environmental issues are seldom included in these sets of final values (Breemhaar et al., 1995, pxvii-xix). This line of argument opens the way for research into social mechanisms for constructing sets of final values in combination with social differentiation of lifestyles (see e.g. Schwarz and Thompson, 1990; Motivation, 1999), an approach scarcely followed in social environmental research up to now.

At this point, we conclude that there is public support for addressing climate change, at least at the level of attitudes. However, the public doesn't respond to the greenhouse issue as such but to environmental issues in general. Up to now, the climate issue plays scarcely any role in day-to-day behaviour and there are clear signals that it isn't given a high priority. There is little to be expected from pleas for sober lifestyles.

Producers

In the 1990s impressive improvements occurred in the general environmental performance of producers. In the late 1980s Dutch environmental policy set ambitious targets for the period 1990-2010; in some respects demanding an 80 per cent reduction in an even shorter period than that discussed in this book for CO_2. Producers were addressed with a mixture of three policy strategies: a central approach using modern forms of coercion and incentives, an interactive approach and internalization, and a self-regulation approach (Vermeulen, 2000a). In 1998 the Dutch government and producers together published a book entitled *The Silent Revolution* singing the praises of their 'co-operative efforts in improving the environment'. In Chapter 1 we said that these targets will be met (Ministerie van VROM/VNO-NCW, 1998b, p160-161; RIVM, 1999). This success is supported by the institutional changes in businesses during the 1990s. We are witnessing a continuing diffusion of institutional change. The number of companies that have integrated environmental management into the business process or are in the process of doing so has grown from 37 per cent in 1997 to 49 per cent in 1999 (KPMG Milieu/NIPO, 1999: 24; see also KPMG Milieu/IVA, 1996). More recently, we have seen rapid growth in the number of firms with ISO 14001-certified environmental management systems (ISO, 2000).

We should be aware that, by implementing environmental management systems, companies in theory put in place the institutional requirements to meet the governmental and societal demands in all areas of environmental improvement, including climate change. However, during the 1990s climate change and energy-efficiency policies have followed different paths, being instigated by different government departments and employing different policy instruments. During the 1990s the Minister of Economic Affairs, who is responsible for energy policy, established a separate forum for consultation with the aim of reaching voluntary agreements on improvements in energy efficiency. As we discussed in the first section of this chapter, this approach has proved effective. Despite low energy prices, companies have met the agreed targets. Research shows that they are more motivated and have improved their organizational capacity to increase their energy efficiency, and energy efficiency has taken a more prominent place in their decision-making procedures (Glasbergen et al., 1997; Glasbergen, 1998).

The first generation of energy-efficiency agreements was mainly concerned with optimization of production processes, application of combined heat and power and energy conservation. They did not cover the use of renewable energy. Natural gas and electricity are simply taken from the national grid. Large industrial firms deserve special attention as large-user tariffs are an interesting issue to address, especially since the European energy market is in the process of liberalization. In 2000 only a handful of businesses used renewable energy, like photovoltaic cells, wind energy or electricity with green certificates (examples are the national rail company NS, Van Melle, a manufacturer of sweets, and Siemens, as producer of wind turbines). In doing so they have faced opposition. For instance, Van Melle could not get local permits to place wind turbines at any of its sites in The Netherlands.

Recent research on 'cleaner production' shows us what types of measure are being applied by businesses. Depending on the type of measure, between 25 per cent and 65 per cent of companies have applied specific measures. Most of these involve a reduction in the use of materials and adaptations in production processes, but there have also been energy-related measures (between 60 and 65 per cent of the companies) (KPMG Milieu/NIPO, 1999).

There is little specific information about businesses' attitudes towards the climate change issue and climate change abatement. A study from 1995 involving over 500 business decision makers in The Netherlands (Research and Marketing: 1995) shows that about one in four were strongly committed to their environmental performance and one in five showed a strong commitment to climate change issues. However, other (more traditional) environmental issues are regarded as far more important. In general, the level of knowledge and perceptions of climate change issues differ little from those of the general public (see also the previous section on public opinion). Only one in three expected effects on society in the short term. Their main source of information was mass media, but two in five said they generally ignored articles in newspapers or programmes on television about this issue.

Nevertheless, business decision makers generally see industry as one of the main perpetrators (85 per cent) and it is generally accepted that industry has a leading role to play in abating climate change (89 per cent), while also expressing great confidence in the (technological) capacity to resolve the problem. The main barriers to solving the climate change problem are felt to be economic interests, conflicting policy interests and the inadequate pace of technological development. About two-thirds of those surveyed stated that their own company should take initiatives. Regrettably, the study didn't look in more detail into the willingness of companies to act themselves and what they did in practice. A more recent study (Van der Woerd et al., 2000) reported a tendency amongst multinationals in the US and Europe to move towards a more proactive approach towards the climate problem.

Another study focuses more specifically on investment behaviour by firms (Gillissen et al., 1995) and describes barriers to diffusion, the investment criteria that are applied and policy preferences. Their analysis shows that 20-30 per cent of the technological potential for energy conservation is unknown to industry, even in the case of profitable technologies. Perceived impacts on production quality and continuity (possibly due to 'start-up' problems of new technologies) are important barriers and it was found that firms' perceptions of profitability differ significantly from profitability calculated in policy studies (Gillissen et al., 1995, p89-91).

Our observation here can be that there are visible changes in modes of production and in the integration of environmental management approaches. There is some difference between large (internationally-oriented) firms and SMEs. Multinationals like Shell and BP-Amoco are currently shifting their position and are leaving traditional lobbying organizations against climate policies and moving towards a more proactive position. Energy-efficiency is receiving growing attention, but is often not yet integrated in environmental management. Even when it is, it has to compete with other cost-reducing (or liability-reducing) environmental issues. Scarcely any attention is paid to 'renewable energy'. Chapters 4 and 7 will discuss possible routes for the acceleration of dematerialization and technological development.

Consumers

Consumer behaviour is relevant for CO_2 emissions. Energy use in households for heating, electricity and use of appliances has been the subject of energy policy for nearly a quarter of a century now. The growing mobility of consumers adds substantially to the volume of CO_2 emissions they generate. And, more recently, attention has also shifted to the entire volume of consumed goods. Every product bought may use energy during domestic use, but also includes energy used for its production and disposal. This indirect energy requirement of products and services consumed actually exceeds the energy requirement for domestic use of electricity, heating and use of petrol for cars (Biesiot and Moll, 1995, p43).

Governments have been attempting to influence domestic use of electricity and heating and the use of petrol for cars since the mid 1970s, when The Netherlands was boycotted by Arab oil producers. Many lessons have been learned, but it would take too long to discuss them all here (see Piers Consultancy, 1992). The best way to see whether policies were effective is to look at physical effects and market information. These show that the use of electricity rose during the 1990s at an average of 3.4 per cent per year. The use of natural gas is declining due to the successful promotion of insulation and high-efficiency central-heating appliances (Brezet, 1994).

In the field of mobility we are far from a trend break: the number of kilometres driven has been growing at 2 per cent per year, although the rate of increase seems to be slowing. That is explained by a dramatic growth of congestion on the roads. We see a gradual growth in the use of public transport (3.5 per cent per year), but it still only represents a small proportion of the volume of private transport (about 1:7 to 8) (RIVM, 1998b; CBS Statline).

The other aspect being considered here, the entire volume of consumed goods, is worth looking at in more detail. Biesiot and Moll studied the energy requirements of consumption patterns. This study shows a strong relation between income levels and total energy requirements: households with an income of € 10,000 on average use twice as much energy as households with an income of € 5,000. But even more interestingly, we see large differences within all income categories: those who use a lot of energy use twice as much as the thrifty users within the same income group (Biesiot and Moll, 1995, p48). Although this may be largely explained by physical factors (quality of the houses) and situational factors (distance to work), choices in consumption present quite significant opportunities for greenhouse gas reduction by means of conscious use of products and services with a low energy requirement. Another study shows that motorists themselves believe that 20 per cent of their car use is avoidable (Tertoolen, 1994, p274). A large-scale experiment in The Netherlands showed that with intensive efforts to raise awareness, consumers could combine increased wealth with a reduction of total energy requirements by 30-40 per cent (Schmidt and Postma, 1999).

Results in real-life situations give far less reason for hope. Tertoolen tried to influence car users' behaviour through persuasion with arguments concerning the collective and personal disadvantages on the one hand, and by improving self-monitoring, feedback and commitment on the other. He concluded that car users will not stop voluntarily because car use is predominantly linked with feelings of independence, ease and comfort (Tertoolen, 1994, p278).

When we look at purchasing behaviour, we can use information from a longitudinal study of environment-related behaviour (Couvert and Reuling, 2000, p108-151). It showed that 10 per cent of the population often takes the environment into account in purchasing decisions and 20 per cent do so sometimes. The others just choose on functional or economic grounds. In the period 1993-1999, these levels were generally stable for most types of product in the study. There was a minor tendency among high-income groups and environmentally-oriented individuals to improve their 'performance'. In the last year, low-income groups and people who are indifferent towards the environment tend to drop out with respect to purchasing products with less effect on environment and separation of waste.

Here again it is most instructive to look at information available about the effects at the end of the production and consumption chain, or to put it another way the end of the *information –> decision –> behaviour* chain. The study discussed above is based on what people say they do. The best proof of behaviour is to look at what is actually being sold. Looking at purchasing behaviour, we see only very small market shares for eco-labelled products. As regards food (representing half of the total indirect energy requirement of households (Biesiot and Moll, 1995, p43), the market share of milk produced under the European EKO label for organic food products grew from 0.6 per cent in 1998 to 1.5 per cent in 1999 after supermarkets changed their policies. Comparable figures for other food products show us three things (Vermeulen, 2000):

1. far more people say in surveys that they buy organic food than actually do;
2. actual market shares for organic food are very small. But there is a small glimmer of hope;
3. changes in policies can have large effects in market niches.

The central question here is whether addressing consumers with both information about the relative environmental impact of alternative products and improving the supply of the environmentally improved goods will bring about trend breaks. The non-governmental organization Global Action Plan employs a method of information transfer, self-monitoring and social stimulation in assisting small groups of citizens with changing their behaviour. These groups are called eco-teams. Drawing on evaluations of eco-team behaviour, they claim to be able to realize a 25 per cent reduction of CO_2 emissions (GAP, 1999). Others like Aarts and Breemhaar, whose arguments we discussed in the section on public opinion, are far less optimistic.

These observations have recently led to what might be called a withdrawal from addressing citizens as conscious consumers. The Dutch National Advisory Council for the Environment recently stated in its advisory opinion on 'sustainable consumption' that it is unwise to expect consumers to buy more expensive product alternatives with better environmental performance. Urging them to do so would imply an overly moralizing role for government. Instead, emphasis should be placed on addressing producers (Raad voor het Milieubeheer, 1996, p79, p84, p85).

Our main conclusions from this section are that we are seeing a shift from an approach where consumers are addressed mainly with respect to energy conservation through their behaviour (heating, lighting, buying low-energy household applications) to one that addresses consumption as a whole. Recent environmental policy has addressed citizens as consumers, but with marginal results. Promoting sober lifestyles tends to be ineffective as it is too contrary to dominant values. Willingness to make voluntary financial sacrifices is small. Rising abatement costs might collide with low acceptance within segments of the population (related to level of education and lifestyles). At the same time, government's attention for the consumers' role may need to be reinforced.

Reflection: main dilemmas

Surveying the research discussed above, we can see a widespread recognition and acceptance of climate change as a problem that has to be dealt with. This applies to both the level of policy-making and the level of society (stakeholders and public opinion) in The Netherlands. The 1990s also highlighted a willingness to set quantitative policy targets, and at the local level even quite substantial ones. However, they address the short term and involve a step-by-step approach. Because of the difficulties experienced in changing energy-related behaviour, targets have been moderated and even those are hard to achieve.

In addition, looking at the actual behaviour of the relevant societal actors, we see a gap between a willingness to act at the political level and the slow change in actual behaviour of target groups or, in other respects, the absence of the climate change as an issue in their daily lives.

We have consciously focused on the situation in the 1990s to show the contrast between the long-term images shown in the previous chapter and the current situation in Dutch society. Of course this only gives us a picture of attitudes and actual behaviour for a specific and short period, and shows a difficult struggle with very modest policy targets. Over the longer term, perceptions, attitudes and daily routines may prove to be less irreversible than suggested by these observations. However, in our quest for societal barriers in the development towards arriving at the images presented in Chapter 2, we are better off taking into account these constraints and finding ways of dealing with them. Even if we assume that the climate change problem is recognized, resistance among different groups in society based on considerations of cost-effectiveness and competing interests and goals will have to be faced. On the other hand, in our open, non-directive society, long-term and substantial climate change policy will inevitably require co-operation by these different groups in society. Knowing this, we can identify various obstacles to securing public willingness to act:

- a substantial segment of public opinion doesn't support major interventions (relation with income levels, lifestyles and specific political cultures);
- reallocation of administrative competences implies that national governments may have political targets but will increasingly depend on others in attaining these goals;
- in local and regional politics substantive action towards greenhouse gas reduction has to compete with more urgent and inescapable local priorities. This in a context where effects of climate change are barely perceptible and local authorities will rarely be the key actors in combating climate change;
- for many (but not all) of the societal groups involved, other 'basic' values than environmental consciousness construct the initial motives behind choices and orientations of daily behaviour. Conduct relating to greenhouse gases will have to be pushed into these 'basic values-driven' behavioural routines or will have to produce co-benefits;

- there also appear to be some obstacles within industry. There are clear tendencies towards improvement of environmental performance, but for this group of stakeholders greenhouse issues have to compete with other environmental and socio-ethical issues, often involving more direct responsibilities.

The following chapters will explore the most relevant routes for social transformation and identify necessary trend breaks in social institutions, cultures and practices. The authors will adopt different academic perspectives and focus on various sectors of society. The obstacles discussed above confront us with a number of difficult choices one might not want to make between alternative approaches: seven dilemmas in defining strategies for achieving a climate-neutral society. These dilemmas are presented next.

1. *The dilemma of a 'small number of decision makers' trajectory versus a 'decentralized, multiple and diffuse decisions' trajectory*
Do we prefer to realize an 80 per cent reduction by means of a 'central-technological' trajectory, choosing to apply large-scale technological solutions and clean production, or with a trajectory involving as many societal actors as possible, all asked to do their bit to solve the problem in their roles as consumer, motorist, producer, investor, decision maker etc? The second route may be very sensible, opening up as much creativity in society as possible, but it requires sufficient involvement of society. On the other hand, while the 'small number of decision makers' trajectory may very well be feasible (e.g. to build enormous wind parks in the North Sea), it may reinforce the weak public support and may even contribute to the growing distance between politics and people.

2. *The dilemma of 'acting now' versus 'delayed response'*
This dilemma refers to the timing of climate policies. Long-term targets will determine the necessary efforts. According to some, if an 80 per cent emission reduction is to be achieved we have to act immediately. The priority needs to be on reducing emissions and stimulating technological developments in the coming 15-20 years. 'Optimists', on the other hand, believe that a delayed response is appropriate. Future technological progress will enable rapid, cheap and drastic emission reductions in the future, so there is no reason to intensify climate policies right away.

3. *The dilemma of 'focusing on climate change' versus 'pursuing sustainability'*
The public does not perceive climate change as a separate problem. This non-exclusivity of the climate change problem applies equally to many of the technological solutions and policy options. Pushing through a -80 per cent transition may require separate and selective attention in the form of special campaigns and policies on climate change, especially when target groups (still) scarcely acknowledge the problem or the sacrifices needed. Separate public campaigns, separate routes to solutions, clash with existing preferences for attacking climate change in integrated environmental, or preferably sustainable development, approaches.

4. *The dilemma of 'demystification' versus 'need for increased awareness of change impacts'*
   Research into environmental attitudes shows a gradual demystification of apocalyptic messages concerning the environment. In the context of 80 per cent reduction of greenhouse gases, the lack of support for climate change issues at the level of genuine action calls for campaigns designed to increase understanding of climate change and the need to abate it. In that case, apocalyptic messages tend to be counter-productive.

5. *The dilemma of 'central steering via direct regulation and market incentives' versus 'co-production'*
   Regulation and market incentives in theory seem to be effective if enforcement is well organized and incentives are substantial enough. However, due to the administrative burden and reallocative effects, both forms of central steering have serious difficulties in emerging from the political decision-making process unscathed. Participatory approaches in developing climate change policies (co-production) may be the solution, but they require clear politically determined targets and a context of direct regulation and market incentives, discouraging free riders.

6. *The dilemma of 'transition through co-benefits' versus 'inevitability of a special climate change approach'*
   Opting for solutions that are attractive thanks to the co-benefits (simultaneous benefits of solutions, like cost reduction, functional qualities, comfort, safety etc.) makes it possible to ignore lack of societal support for climate change policies alone. However, although this may help in the first stages on the way to an 80 per cent reduction, it will probably not be sufficient in itself. A special climate change approach may be inevitable.

7. *The dilemma of 'decentralization and Europeanization' versus 'the need for national direction of the transition'*
   Achieving an 80 per cent reduction will not only require clear choices in technological fields and modes of enforcing their implementation, but also constant monitoring, activation and revision of policies. We tend to assign this role to national governments, but due to two bi-directional processes, decentralization and Europeanization, European national governments have fewer tools in their own hands and depend more and more on the co-operation of others.

Too many dilemmas may make it hard to see whether we will ever be able to make the transitions described in the previous chapter. However, the practices, policy initiatives, policies under construction, partial support in the various groups in society discussed above could equally be seen as a breeding ground for initiatives that, given fertile growing conditions, will flourish in due course. The following chapters will explore the necessary *societal* trend breaks, consciously bearing in mind the tensions between ambitions and recalcitrant societal practices.

References

Aarts, W., *De Status van soberheid*, Amsterdam, 1999.

Aarts, W., J. Goudsblom, K. Schmidt and F. Spier, *Towards a morality of modernity*, Amsterdam School for Social Science Research, Amsterdam, 1995.

Algemene Energie Raad, *Advies Duurzame Energie*, Den Haag, 1999.

Becker, J.W. et al., *Publieke opinie en milieu*, SCP, VUGA, 1996.

Biesiot, W. and H.C. Moll, *Reduction of CO_2 emissions by lifestyle changes.* NRP Global Change, report nr. 410 100 108, Bilthoven, 1995.

Breemhaar, B, W.A.C. van Gool, P. Ester and C.J.H. Midden, *Leefstijl en huishoudelijke energieconsumptie*, NRP Global Change, report nr. 410 100 109, Bilthoven, 1995.

Brezet, H., *Van prototype tot standaard; De diffusie van energiebesparende technologie*, Rotterdam, 1994.

Burger, H., *Evaluatie Klimaatverbond*, ECN Petten, 1999.

Couvert, E. en A. Reuling, *Milieugedragsmonitor IX*, Amsterdam, NIPO, 1999.

Couvert, E. en A. Reuling, *Milieugedragsmonitor X*, Amsterdam, NIPO, 2000.

Ester, P. And H. Vinken, *Sustainability and the cultural factor: Results from the Dutch GOES MASS PUBLIC MODULE*, Globus, Tilburg, Dutch National Research Programme on Global Air Pollution and Climate Change, Report nr. 410 200 048, 2000.

GAP, *Het Ecoteam Programma, een meetbare aanvulling op het overheidsinstrumentarium voor het kosteneffectief bereiken van de milieudoelstellingen*, Den Haag, 1999.

Gillissen, M., H. Opschoor, J. Farla, K. Blok, *Energy conservation and investment of firms*, NRP Global Change, report nr. 410 100 102, Bilthoven, 1995.

Glasbergen, P. et al., *Afspraken werken, evaluatie meerjarenafspraken over energie-efficiency*, Utrecht, 1997.

Glasbergen, P., Partnership as a learning process. In: Glasbergen, P. (ed.), *Co-operative environmental management: public-private agreements as a policy strategy*, Kluwer, Dordrecht, 1998.

Groot, H.L.F. de, E.T. Verhoef en P. Nijkamp, *Energy saving by firms: decision making, barriers and policies*, VU Amsterdam, Amsterdam: Tinbergen Instituut, 1999.

Hoefnagel, R. et al., *Milieurelevant consumentengedrag*, SCP, Den Haag, 1996.

ISO, *The ISO Survey of ISO 9000 and ISO 14000 certificates*, ISO, Geneve, 1998.

KPMG Milieu / IVA, *Evaluatie Bedrijfsmilieuzorgsystemen 1996*, KPMG Milieu, DenHaag/Tilburg, 1996.

KMPG Milieu / NIPO, *Schoner produceren in Nederland 1999*, Den Haag, Ministerie VROM, 1999.

Meijders, A., *Climate Change and Changing attitudes, Effect of negative emotion on information processing*, Eindhoven University of Technology, Eindhoven, 1998.

Midden, C.J.M. en G.C. Bartels, *Consument en milieu*, Houten/Zaventem, Bohn Stafleu van Loghem, 1994.

Ministerie van VROM/VNO-NCW, *De stille revolutie: industrie en overheid werken samen aan een beter milieu*, Distributiecentrum VROM, 1998.

Noorman, K.J. en T. Schoot Uiterkamp, *Green Households, Domestic Consumers, Environment And Sustainablity*, Earthscan, London, 1998.

Piers Consultancy, *Energiebesparing op maat*, Den Haag, 1992.

Raad voor het Milieubeheer, *Duurzame consumptie een reëel perspectief*, Den Haag, 1996.

Rayner, S. and E.L. Malone, *Human Choice and Climate Change – the societal framework* (Volume 1-4), Batelle Press, Columbus, 1998.

Rayner, S. and E.L. Malone (1998) *Why study human choice and climate change?* In: Rayner, S. and E.L. Malone, Human Choice & Climate Change – the societal framework, Batelle Press, Columbus, Volume 1, pp. 265-344.

Rijksinstituut voor Volksgezondheid en Milieu (RIVM), *Milieubalans 1998*, Alphen aan den Rijn, 1998a.

Rijksinstituut voor Volksgezondheid en Milieu (RIVM), *Achtergronden bij Milieubalans 1998*, http://www.rivm.nl.., 1998b.

Rijksinstituut voor Volksgezondheid en Milieu (RIVM), *Milieubalans 1999*, Alphen aan den Rijn, 1999a.

Rijksinstituut voor Volksgezondheid en Milieu (RIVM), *Achtergronden bij Milieubalans 1999*, http://www.rivm.nl., 1999b.

Rijksinstituut voor Volksgezondheid en Milieu (RIVM), *Milieuverkenning 5 2000-2030*, Alphen aan den Rijn, 2000.

R&M Research and Marketing, *Rapport van een kwantitatief onderzoek naar kennis en houding van beslissers in het bedrijfsleven t.a.v. de klimaatproblematiek*, Heerlen, 1995.

Schmidt, T. en A.D. Postma, *Minder energiegebruik door een andere leefstijl? Project Perspectief, Eindrapportage*, CEA, Den Haag, 1999.

Schwarz, M. and M. Thompson, *Divided we stand: redefining politics, technology and social choice.* New York: Harvester Wheatsheaf, 1990.

Slob, A. et al., *Trendanalyse Consumptie en Milieu*, TNO-STB, Apeldoorn, 1996.

Sluijs, J. van der, *Anchoring amid uncertainty. On the management of uncertainties in risk assessment of anthropogenic climate change*, PhD thesis, Utrecht University, 1997.

Sociaal Economische raad, *Advies Uitvoeringsnota klimaatbeleid deel I*, SER publicatie 99/14, Den Haag, 1999.

Steg, E.M., *Verspilde energie?*, SCP Cahier 156, Den Haag, 1999.

Tertoolen, G. , *Uit eigen beweging ...?!*, Proefschrift Universiteit Utrecht, Utrecht, 1994.

Thompson, M. and S. Rayner, *Cultural Discourses* In: Rayner, S. and E.L. Malone, Human Choice & Climate Change – the societal framework, Batelle Press, Columbus, Volume 1, 1998, pp. 265-344.

Veen, H.J.C. van der, en J.L. Peschar, *Aanvaardbaarheid en politieke haalbaarheid van energiebesparende maatregelen*, De Lier, 1995.

Vermeulen, W.J.V., *De weerbarstige consument*, In: P.P.J. Driessen en P. Glasbergen (red.), Milieu, samenleving en beleid, Elsevier bedrijfsinformatie bv, Den Haag, 2000, pp. 353 - 373.

VROM-Raad, Transitie naar koolsofarme energiehuishouding. Advies ten behoeve van de uitvoeringsnota klimaatbeleid, Den Haag, 1998.

Waals, J.F.M. van der, S. Joosen, B.P. van Geleuken, M.C. Groenenberg, M. Kneepkens en W.J.V. Vermeulen, *CO$_2$-Reduction In Building Locations: A Survey And Three Case Studies About The Role Of Options For Co$_2$ Reduction In Planning Processes*, Utrecht University, Dutch National Research Programme on Global Air Pollution and Climate Change, Report nr. 410 200 036, 1999.

Waals, J.F.M. van der, S.M.J. Vermeulen, W.J.V., Vermeulen, P.Glasbergen and P. Hooimeijer, *Energiebesparing en stedelijke herstructurering, een beleidswetenschappelijke analyse*, Utrecht, DGVH/Nethur partnership 10, 2000.

Woerd, K.F. van der et al., Diverging Business Strategies towards Climate Change. A USA-Europe comparison for four sectors of industry, IVM, NRP report no. 410 200 052, Bilthoven, 2000.

Chapter 4

Technological change and innovation for climate protection: the governance challenge

Maarten Arentsen, René Kemp and Esther Luiten

In this chapter...

... the role that climate-oriented technology policy should play in reducing CO_2 emissions by 80 per cent towards 2050 will be addressed. A lot is expected from technology as regards achieving such radical emission reductions. After Chapter 2, which described two possible ways to realize drastic emission reductions, and the analysis of the current situation in Chapter 3, this chapter considers how to realize trend breaks and orchestrate the transition towards a climate-neutral society from a technology perspective.

These questions will be explored with the help of theories of technology dynamics. Contrary to neo-classical economic theory and engineering-oriented innovation approaches that view technology as a transformer, technology dynamics assumes there are patterns to techno-logical dynamics and development. It tries to understand how and why these patterns emerge, evolve and change. A fundamental assumption is that technical change and the social envi-ronment co-evolve: technical change is shaped by human needs and the exercise of human agency, while the social environment is shaped by technology. Through technology we change, a view which suggests that technology is more than a mere transformer.

Introduction

The challenge of climate change demands radical changes in technology and new directions in innovation to satisfy the technological needs of the climate-neutral economy. According to the IPCC analysis, the Western economies may have to reduce greenhouse emissions by 80 per cent in 50 years, a very short period on the time-scale of technological development. The future world with 80 per cent fewer CO_2 emissions sketched in the A and B visions in Chapter 2 of this book clearly indicates the technologies needed in such a world. The future world calls for tech-nologies that differ in almost every respect from the technologies currently used in all sectors of the economy.[1] The question facing modern industrialized society is, therefore, how to initiate such technological changes, whether they should be incremental or radical, and what the role and contribution of the actors involved,

---

[1] It should be noted that these radical changes are needed throughout the economy, including vital sectors currently dominated by fossil fuels, such as heavy industry, the energy sector, construction, transport and agriculture. See, for instance, the list of changes needed in industrial production, agriculture, transport, energy supply and consumption in Chapter 3.

in particular government, is or should be. Although governments are not the only actors charged with seeing through the change process, in industrialized society they bear a special responsibility for guiding future developments. Only democratic governments possess democratically-legitimated power to make authorized decisions in this regard, because of the normative nature and content of such decisions. But governments cannot and should not work alone in the change process towards a climate-neutral economy.

Our analyses will show that desired climate-neutral technologies cannot be ordered or chosen as large or small scale technological options (compare dilemma 1). Radical change emerges from constant and continued search for better technologies. Furthermore, technological development is by definition a multi-actor process and driven by the many goals and ambitions of the actors involved and not by the desire to reduce the climate change problem. What can and should be done, therefore, is to work on the internalization of climate protection needs in these multi-actor processes. In this way climate protection might become a new value orientation of the many, many actors carrying innovation and a new guiding principle in the search for the next technological steps. Each new step might incubate a radical change and, therefore, it makes no sense to wait and hope for radical change in the future (compare dilemma 2). Internalization of climate change needs in processes of technological innovation and change is a long-term process by definition and requires diversified governmental actions and a rich policy repertoire both at the national and the international level (compare dilemma 5). Though we might know the kind of climate-neutral technologies needed, the process to initiate and to develop these kind of desired technologies cannot be managed top down. The process itself is by definition highly differentiated, quite often of a global scale with many, many different actors involved, each influencing the course and direction the process takes in terms of new technological steps. Policies, therefore, are constantly facing the dilemma to focus explicitly on climate change needs or to tackle these needs more indirectly as co-benefit (compare dilemma 6). What is best cannot be prescribed, but should be decided by a clear analysis of the technical and non-technical inertia involved in processes of socio-technical change. In the context of this chapter it is only possible to highlight some of the general aspects of socio-technical change processes from an evolutionary point of view.

The evolutionary view of technical change is based on the idea of innovation as part of systems and trajectories. With the help of this idea it is possible:

1. to analyse how and why technology predominantly changes along fairly incremental paths of development and why it is so difficult to initiate more radical types of innovation. In the next section incremental change in technology will be explored in more detail.

2. to explain what types of innovation are needed with a view to reducing CO_2 emissions by 80 per cent in 2050. The A and B visions assume fairly radical innovation of existing technologies, such as a 40 per cent increase in material efficiency in industrial production and consumption. Therefore, next we discuss the implications of these kinds of improvements in terms of system innovation and transitional change.

3. to identify new points of intervention and types of policy. In particular, the evolutionary view stresses the need for policy makers to be actively involved in the vari-

ation-selection processes that change technology, e.g. by stimulating the invention of new technology and the selection of climate-friendly technology in the economic process. Based on this, we go into the implications of an evolutionary-based analysis of technological change for technology policy. Our analysis will show a need to design and implement short-term policy strategies from a longer-term transitional perspective. Finally, the last section of this chapter summarizes the major conclusions of the analysis.

Lessons from history: Why technology tends to change incrementally

One of the key lessons taught by scholars of technology dynamics is the high degree of complexity and the collective nature of technological change processes, which makes them fairly difficult to influence, for instance from a climate change perspective (although not impossible, as will be shown in later sections of this chapter). The scientific understanding of this complexity is still rather fragmented, but literature refers, among other things, to the multi-layered nature of technological change and to the solid embedding of technology in society as factors contributing to complexity. Technologies are part of functional systems (food production, energy distribution etc.) that have evolved, and their evolution has given rise to beliefs, institutions, infrastructures and interests that guide the actions and the thinking of the actors. So technical change does not occur in leaps. A new technological invention initially tends to develop step-by-step over time with small improvements of its technical and economic performance. This is because technological development builds cumulatively on past achievements, ideas and knowledge. Past achievements are not only visible in the technologies and technology systems around us today, but also in scientific laws, technological principles, science and engineering practices and search routines through the training and education of new generations of scientists and engineers. Technology builds on this cumulative knowledge. Individual innovations also depend on complementary technologies for their production and use. In this way, technologies and technology systems build and develop their own irreversible patterns of development which focus and narrow the way we search for new and better-performing technologies (Rosenberg, 1982). In technical change, as in human life, there are path-dependencies ('lock-in').

If we repackage the empirical materials (see Boxes in this section) in more theoretical concepts, we might say that technological changes take place within a material context and a cognitive and normative framework. Each societal function (such as production, transport and mobility, heating and illumination) is satisfied within a specific socio-technical system. For our transport and mobility needs we developed car-based and truck-based transportation systems, for our production needs we developed specific industrial systems (iron and steel, agriculture, chemistry) and for our power needs we built electricity systems. The systems developed in specific ways and according to specific patterns, while technological change within these systems was guided by the cognitive and normative framework, which is known as a 'technological regime' (Nelson and Winter, 1977; Kemp et al, 1998). The regime consists of beliefs and rules that guide the actions of actors.

Box 4.1

Patterned nature of technological change

One of the major innovations in paper production in the 20th century has been the introduction of shoe-press technology (1980). The shoe press is regarded as one of the major historical breakthrough technologies regarding energy efficiency in the paper industry. Shoe-press technology was an innovative component for the pressing section, one of the important components of the conventional process of paper production. Although the technology brought about a substantial improvement in the production capacity of paper machines, it was merely an incremental improvement if one evaluates this technology from a system perspective. This technological change fitted in with the existing regime search heuristic for improving paper production: increase dewatering so that the machine speed could be enhanced and the annual production capacity could be increased. That the innovative technology also saved energy was only of secondary importance to both developing and implementing the technology. The subsequent development of impulse technology, a new pressing technology, which builds upon the shoe press, can be seen as a continuation, and thus a further optimization, of the existing regime. Impulse technology is designed to further optimize dewatering, increase machine speed and reduce capital intensity. These measures reflect the most important need of paper manufacturers. The paper industry is highly locked in to optimizing the existing regime. Within that regime, quite radical technologies may be introduced (like the shoe press, which was a real break-through from an engineering point of view, although it further stabilized the regime of optimizing dewatering and thus increasing production capacity).

The existing technologies and rules exert a significant influence on technical change. When the use of new technologies requires important changes at the point of use and upstream, the development will be slow. Two technologies that illustrate this are strip-casting technology and smelting-reduction technology, two energy-efficient technologies in the steel industry. Strip-casting technology, for example, is based on an old idea which was already suggested by Bessemer around 1860. It was more than 120 years, the period 1980-1985, before economic actors in a number of small networks seriously pursued the development of this technology. Whereas the idea (and the huge advantages) were known for decades, two preceding casting technologies had to materialize before strip casting was generally considered by steel makers as 'the next step to take' in casting technology. Only then had the need for thinner casting become clear for categories of steel producers that didn't exist during the days of Bessemer. The steel industry had changed; stainless steel producers and mini-mill operators had arrived on the scene. In addition, more basic knowledge was gained in areas that turned out to be suitable for bringing strip casting towards commercialization. A niche emerged in which the innovation could emerge (Luiten, 2001).

Box 4.2
The emergence of the modern electricity system

The invention of electricity is perceived as one of the major achievements of mankind. Thomas Edison is generally recognized as the founding father of the modern electricity system with his groundbreaking system's approach to electricity. The secrets untangled by Thomas Edison and his successors enabled the construction of the central station electricity system which evolved into the system that we have today: a rather homogeneous technology consisting of large power plants interconnected by a system of high- and low-voltage cables transmitting the (transformed) electricity directly to consumers. Electricity production, transmission and consumption are well integrated into a system consisting of a large set of interconnected technological components (Hughes, 1983).

The basic technological principles guiding this process were formulated in the 1920s and have not changed since (op.cit. p. 370). The electricity system built on these technological principles is rather homogeneous throughout the industrial world and the technical configurations differ only in the mix of energy resources (fossils, hydro, nuclear) used for power generation. More than a century of engineering practice, learning capacities, skills, science and engineering has been devoted to and invested in the central station electricity system as we know it today.

System change from within: hybridization of power-generation technology
Steam turbines long dominated thermal, fossil-based power generation. Gas turbines were developed because they could satisfy the increased need for flexibility in (peak) load demands in the central system. In a next step, gas turbines were also used in combination with steam turbines for base load power generation. The combined cycle of gas and steam turbines significantly improved the efficiency rate of energy conversion. At the same time, however, gas turbine technology paved the way for two important developments: the combined production of heat and power (CHP) and the decrease of scale of power-generating technology. Both CHP and the decrease of scale initiated technological developments that threatened the domination of the large-scale central station electricity system. In the 1990s CHP endangered the central co-ordination of the Dutch electricity system and the decrease of scale gave way to the development of small-scale stand-alone power systems able to satisfy heat and power demands of individual households. These new technological developments, deviating from the dominant ones in the large-scale central station electricity system, have been supported by institutional changes in the organization of national power systems during the 1990s. Liberalization of electricity markets in this way dramatically changed the perspective and potential of smaller-scale power-generation technology. In decentralized competition and price-based electricity markets, small-scale and flexible generation capacity turns out to be an advantage and a far less risky investment than for instance a large-scale nuclear power plant.

Box 4.3
System change initiated outside the dominant system

The gas turbine illustrates how new technological routes may start within a dominant system. The gas turbine was a hybrid technology incorporating the potential to meet the needs of the current system, but at the same time incorporating a potential to develop a new technological route deviating from the current system. Apart from within the system, system change can also start outside the current system. Here too the modern electricity system provides a good example. The climate change problem calls for a significant increase of renewables as an energy resource for thermal-based electricity systems (see both visions in Chapter 2). Despite the early recognition of the need to develop alternative renewable-based technologies for power generation, for a long time in The Netherlands their development was rather hesitant. Politically, the need for their development was first expressed in the aftermath of the first energy crisis in the early 1970s. The political ambitions were almost totally ignored by the electricity system, which continued innovation along the predominantly large-scale technological paths of the central station electricity system. Within the system, innovation efforts were devoted to continuing to use fossil energy resources and to developing nuclear-based power plants. The development of renewables took place almost entirely outside the electricity system and in The Netherlands it was initiated by new generations of passionate scientists and engineers determined to relieve the environmental hazards caused by dominant fossil-based power-generating technologies. These early idealists were for a long time ignored and not taken seriously by the dominant electricity system and the government, because the dominant thinking in those days was that renewables could never become a serious alternative energy resource for power generation. Denmark took other decisions at that time and these led to the development of wind-turbine technology, giving the country a leading global position in that technology.

Only recently have the prospects for renewables as a power-generating resources changed, due to political ambition expressed both nationally and by the European Union. For that reason, renewables started entering the Dutch electricity system. Their current share in power generation is rather limited and lags behind the political ambition. But what if The Netherlands, like the Danes, had redirected R&D resources much earlier to renewables? The dominant thinking at that time did not allow for such a change, and policies at that time were not strong enough.

Actors invest in technological change according to 'rules' or 'heuristics'. These consist of the engineering consensus about the relevant problems and approaches to solving them, assumptions about what users want, ideas about how the system should be organized, and the roles of actors. Actual decisions regarding technological change that are taken at the micro-level are strongly influenced by the rules of a regime and not just the rules of the market. If a regime is stable, technological change can become relatively autonomous in the sense of predictable; actors look for technological change in a familiar direction (e.g. improving car technology rather than changing car-based transport itself).

The path-dependent nature of technological change is not just a matter of engineering consensus about the relevant problems and approaches to solving them, but also depends on the value of investments in particular technology systems and the costs of building new systems. The examples of technological change within mature manufacturing industries such as pulp and paper and iron and steel illustrate this. The existing layout of capital assets is used as the starting point for technological change within these 'sector' regimes. The main production route is to optimize the existing system to deliver products at the lowest possible cost. This example illustrates how and why existing and dominant regimes tend to induce 'incremental' innovation in the majority of socio-technical systems. Such innovations lead to 'system optimization' rather than 'system innovation' (see also the section on long term technological change below). System innovations occur if society manages to satisfy certain functional needs in a completely new way, i.e. in ways that strongly deviate from the prevailing socio-technical system. In many cases, these types of fundamental change occur through an accumulation of several related innovations together inducing the emergence of a new technological regime. In this way, existing regimes are changed, often with quite a lot of reluctance, for a new set of principles, rules and practices, which will guide further technological change in the system (see Box 4.2 on the electricity system).

Innovation and technological development take time, almost always years or decades, as the example of the shoe-press technology in the paper industry illustrates. This is because technological change and development is not a programmed, or top-down, process. Change processes involve millions of actors and their daily decisions and activities. Neither climate change problems nor other societal needs and demands guide these decisions. At the micro level, decisions are guided by the wide variety of economic, institutional, scientific and engineering rules that form a technological regime. Actors are motivated and guided by the economic logic of the market, the political logic of the bureaucracy or the scientific logic of the laboratory. A variety of incentives, past experiences and future expectations motivate and influence these actors in their decisions and activities that jointly shape technological change. Because of the many actors involved and the rule-guided nature of technical change, technology develops in a step-by-step (incremental) and patterned fashion. Technology tends to change in specific directions, along specific paths of development, which tend to endogenously proceed in certain directions. Changing or redirecting these courses and paths, for instance in climate-friendly directions, proves rather difficult, but not impossible, as will be shown in the following sections of this chapter.

Modulating dynamics: long-term technological change

Incremental movement along the route of technological development will probably not bring about the economy with 80 per cent fewer CO_2 emissions depicted in the A and B visions. Both visions assume the emergence of radical system innovation. System innovation involves changes in technology systems beyond a change in parts or components of technology. It is associated with new linkages, new knowledge, different rules and roles, a new 'logic of appropriateness', and sometimes new organizations. Historical examples of system innovation include the electrification of manufacturing and homes, the development of the central-

Box 4.4
Some examples of system innovation

An example of a 'green' system innovation is industrial symbiosis (industrial ecology): the closing of material streams through the use of waste outputs from one company by another. A (well–publicized) example of industrial symbiosis is to be found in Kalundborg, an industrial area on the Danish coast where material and energy outputs of companies are exchanged. In this area sulphur by-products from a refinery are distributed to a local sulphuric-acid manufacturer and hot water to local greenhouses. Waste heat and steam from the local power station are used by an industrial firm producing enzymes (Cohen-Rosenthal and McGalliard, 1998).

Two other examples of system innovation with potential for climate protection can be mentioned. The first is a hydrogen economy in which the hydrogen is generated in clean ways (for instance, through the use of renewables). The second is integrated mobility, where people use different transport modes (collective and individual, such as a car and bicycle) and transport services are geared to one another. The latter involves not only changes in infrastructure (in the form of P+R stations and special bus lanes) and new technology such as light rail in conurbations, but also major social and organizational change. Such changes include the collective ownership and use of cars (car-sharing and riding), the creation of mobility agencies offering and selling intermodal transport services, the integration of collective transport schemes and the introduction of a transport management system for employees by companies. That system innovations can occur even in a mature industrial sector like the iron and steel industry is illustrated by the possible introduction of smelting-reduction technology. This alternative iron-making technology is rejected by the traditional integrated steel firms (such as Corus) but is not locked out because it is appealing to steel manufacturers that operate mini-mills. It is supported by the existence of small-scale flexible steel-making units, which form a growing 'niche' within the traditional regime of major integrated manufacturers. The existence of this niche allows the technology to be used and benefit from learning effects. The introduction of smelting-reduction technology may enhance the emergence of this new route, which might eventually dramatically change the regime in steel production. This 'system innovation' is, however, not necessarily less energy-intensive than a further incremental optimization of the conventional regime (Luiten, 2001).

ized electricity system, the cluster of synthetic materials innovations and machinery for injection moulding and extrusion. A system innovation of our own time is e-commerce, with the Internet as a new infrastructure for communication and Internet providers as new agents. It should be noted that system innovation does not necessarily imply change in every component or at every level of the 'former' system. System innovation usually consists of a combination of new and old components and may even consist of a novel combination of old components.

The notion of system innovation is highly relevant for the topic of this book (mitigating the climate change problem) because system innovation is one way of achieving environmental benefits and other types of sustainability benefits (in terms of natural capital being preserved, health gains and improved social well-being).

The Dutch DTO programme[2] has identified several sustainable technologies (technology systems) that offer efficiency improvements of considerable magnitude. These include novel protein foods (improvement by a factor of 10-30), precision agriculture (improvement by up to a factor of 50), decentralized production of electricity using renewables and microturbines, underground transport of commodities in pipelines (improvement by a factor of 10 in energy efficiency) and industrial ecology.[3] All these solutions offer benefits for climate protection in the form of a substantial reduction of greenhouse gas emissions, although they are not, apart from renewables, viewed as climate-protection technologies. Their development could be supported from a climate-protection perspective, which requires the co-ordination of climate protection policies with sector policies, an issue we will discuss at the end of this chapter. The time-path of improvement in environmental efficiency for different types of system change is visualized in Figure 4.1.[4]

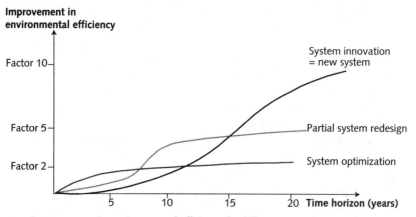

Figure 4.1 *Improvements in environmental efficiency for different types of system change*

The figure shows that in the short term the greatest gains in environmental efficiency can be achieved through system optimization – the retrofitting or environmental upgrading of existing systems. Examples of system optimization include the use of coal scrubbers and flue gas desulphurization (FGD) technologies at coal-fired power plants, the use of energy-efficient manufacturing processes and internal recycling systems in companies, and in the case of products, the shift towards low-solvent paint products. System optimization may involve some radical change and quite costly investments (as in the case of FGD technologies) but does not alter the overall system logic or the paths of development. System innovation (system renewal) and partial system redesign involve wider change, beyond the level of technical parts and components, encompassing the use of new technology, new markets and major organizational change.

---

[2] DTO is the Dutch abbreviation of 'sustainable technology development'.

[3] Descriptions of the technology systems are given in Weaver et al. (1999).

[4] The factors are indicative of the magnitude of the possible improvement in environmental efficiency. The actual improvement will differ for each system innovation. There may be rebound effects that reduce the size of the improvement and the system innovation may create some environmental or social problems of its own, as for example in the case of biotechnology and energy crops.

The distinction between system innovation and system optimization is useful because it forces us to think about the long-term consequences of innovations: whether or not they give rise to or contribute to system innovation or alter the current path of development. An example of such a mapping of innovations (and corresponding policy measures) is provided in Figure 4.2.

Figure 4.2 shows whether the innovations and policy measures to counteract transport problems contribute to system optimization or to system innovation. Personalized public transport, such as dial-a-ride services, and CO_2 policies are believed to contribute to system innovation in the form of intermodal travel in which people combine individual means of transport with collective means. Anti-congestion policies are believed to sustain the current trajectory of motorized passenger transport based on the individual use of cars. Some innovations may be part of both system optimization and system innovation (renewal). An example is urban cars, which may be used in combination with collective forms of transport for long trips or as a second or third household car. It is not uncommon that innovations can be used both within an existing system and within a new system. Very often system change (what we call system innovation) emerges out of existing systems. A good example is the development of the computer regime, which has evolved from the regime when computers were used solely for computing activities and have since found new types of application – for process control, word processing and information communication (through the Internet).

Quite often radical innovations are first used in a 'conservative way' within an existing system before they become an element of the new system. Gas turbines are an example of this. They emerged in the electricity system to meet system needs, but at the same time they opened up new technological options for decentralized power generation outside the central station system (see previous section on incremental technological change).

Transitional changes are thus the outcome of innovations and processes that can be labelled as system optimization or system innovation, although which category the innovation falls into may not be clear from the outset and may change with time. Such transitions occur very slowly. They usually take one or two gener-

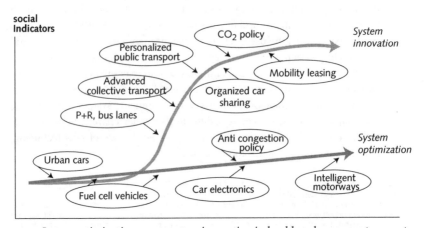

Figure 4.2 *System optimization versus system innovation in land-based passenger transport*

ations (25 to 50 years) and they are not deterministic technological change process-es; that is, they cannot be predicted with any kind of precision. But the change processes that make up the transition can be seen as occurring within one of four different transition phases (see Figure 4.3)

1. a *pre-development phase* of dynamic equilibrium where the status quo does not vis-ibly change;

2. a *take-off phase* where the process of change gets underway because the state of the system begins to shift;

3. an *acceleration phase* where visible structural changes take place through an accu-mulation of socio-cultural, economic, ecological and institutional changes, all of which influence each other. In the acceleration phase there are processes of col-lective learning, diffusion and embedding;

4. a *stabilization phase* where the pace of social change slows and a new dynamic equi-librium is reached.

The distinction between different types of innovation is relevant for (the manage-ment of) environmental technology responses. Environmental protection benefits may be achieved along a route of system optimization or innovation. The first route facilitates quick results, and is usually favoured by the sector and policy mak-ers. The latter route of system innovation offers the prospect of long-term benefits through long-term structural change. Between the two, we have a partial system redesign as an intermediate option or an end option. Some people have made a case for incremental change. For example, Clayton et al. (1999) and Berkhout and Smith (1998) have argued that significant environmental benefits may be achieved through incremental change over an extended period. Whilst this is undoubtedly true, at a certain point in time such changes will run into increasing marginal costs per unit of improvement. If we want to achieve 10- to 50-fold improvements in resource productivity – which according to some are needed in the next 50 years – we need system innovation or technological regime shifts (involving structural change) in addition to the optimization of existing systems or product chains (Ayres and Simonis, 1994; Kemp and Soete, 1992; Rotmans et al., 2000).

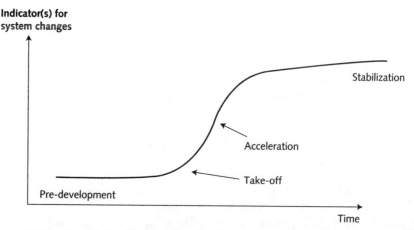

Figure 4.3. *The different phases of a transition (from Rotmans et al., 2000)*

The A and B visions of a low-carbon economy described in this book reflect a range of innovations which are not labelled in terms of 'system optimization' and 'system innovation' but in terms of strategic visions. The strategic visions are made up of innovations and policy measures and changes in the cultural and technological landscape, all of which constitute the building blocks of a new socio-technical system emitting 80 per cent less greenhouse gas in 2050. But it is not really clear how to use these visions for policy and management of the change process. What do they imply for technology policy or any other type of policy? Should we have technology-specific policies or generic policies that change the framework for economic decision-making and technical change? And what does it imply for the government-industry relationship? What is government's role: that of a facilitator, process manager, director or mediator? The next section will turn to these governance questions in more detail.

The governance challenge

The patterns and dynamics of technological change sketched in the previous sections not only indicated the complexities involved but also showed that government is just one of the actors trying to influence the pace and direction of technological change. Change processes are fuelled by millions of decisions of actors and government cannot decisively influence technological change, let alone control and manage such change processes top-down. Initiating the changes needed to bring about the future climate-neutral economy sketched in the A and B visions is therefore basically a challenge of *governance* of socio-technical change processes rather than just a challenge for *government*'s technology policy. Governance refers to decisions and actions of both public and private actors in processes of technological change, whereas government refers to democratically legitimized collective decision-making and policy action. In confronting the climate change challenge, it is of great importance to develop governmental strategies capable of supporting, strengthening and redirecting the governance of the many actors influencing socio-technical change processes in climate-friendly directions.

Two major conclusions from our brief sketch of technology dynamics above could be a starting point for developing these kinds of strategies. These conclusions are:

1. reliance on the autonomy of actors and factors influencing socio-technical change and innovation is not sufficient for a drastic reduction of CO_2 emissions in the next 50 years. This is likely to lead to predominantly incremental changes (system optimization);

2. new technologies and new technological paths (system innovation) do emerge, but the transitional potential of this kind of innovation is hard to identify at the time of their emergence.

Based on these conclusions it is possible to set out some general lines for developing technology-oriented policy strategies.

- First, technology policy should continue to stimulate short-term (incremental) technological innovation, but preferably those innovations incorporating potential for system innovation.

- Second, two types of reference systems should be used to identify the potential for system innovation of incremental innovations: a short-term reference system and a longer-term transitional (system innovation) reference system.
- Third, short-term technology policies should take account not only of the contingencies influencing technological change in the short term but also refer to and fit in with a longer-term transitional perspective of system innovation.

This section sketches some implications for redirecting technological change processes in more climate-friendly directions.

A contingency view of policy

The major lesson that technology studies have to offer for short-term technology policy is that the effect of government intervention depends on the type of technology being considered and the regime in which technological change takes place. Stimulating the development and introduction of photovoltaic energy requires a different government strategy from stimulating the development of energy-efficient technologies in mature industrial sectors. The most important point of departure for developing technology policies is that innovative technologies differ, economic sectors differ, existing technological systems and regimes ruling those systems differ, and hence processes of technological change towards a more sustainable mode of operation differ. There is no standard recipe for directing or accelerating technological change in more climate-friendly directions.

If a government wants to enhance climate-friendly technological developments, it should be clear about the various regimes of technological systems, the actors they can approach and the dynamics of technological change (what drives technological change?). This is the only way to evaluate whether governmental strategies will have an effect, and if so how they will have the greatest effect. Some innovations have a strong enough momentum of their own (for example, strip-casting technology for the steel industry or shoe-press technology for the paper industry) and scarcely need any governmental intervention. Drawing on the particularities of actors, the technologies and the technological regimes as well as the economic sectors should be taken as a point of departure for developing and implementing technology policies. Policies should take into account the whole set of contingencies fuelling technological change.

The choice between technology-specific and generic policies

System innovation requires special support. In appraising innovations eligible for support, one has to ask: does the innovation that produces benefits for climate protection improve the existing regime or offer possibilities for inducing a 'system innovation'? Is there a need for support, and if so how should it be organized?
In defining policies, the following elements should be considered:

1. the improvement in environmental performance of an innovation, e.g. strict CO_2 reduction potential;
2. the contribution of the innovation to system innovation. In what respect does the innovation deviate from existing regimes?

In answering these questions one should consider four aspects:

- to what extent does an innovation physically fit into an existing technological system? Does it replace or complement the existing technological system? Can it be used in both systems?
- does the innovation have a large or small effect on other dominant rules / performance characteristics in the regime? What is the dominant rule in the regime? Is this rule undermined?
- does the innovation offer possibilities for further improvement, for achieving learning economies? Does it have negative effects? Are there ways to deal with them? Is there a need for the innovation or not?
- does the innovation only affect specific firms or entire product chains and/or industrial sectors?

3. are there developments in the economic, social and technological landscape that sustain the innovation? What are these developments and what are their prospects?

4. who is actually involved in developing the innovation? Why? Who will benefit from the innovation? Is there a need for support?

5. which actors can government approach (whom can they influence?) to actually engage in technological change? What government intervention strategy best fits the existing network of actors involved?

Such an assessment of the incremental or radical nature of an innovation's potential can only be tentative. It needs clear understanding of the existing regime as a point of reference. Such an analysis leads to insight into the extent of the changes that are needed to get an innovation introduced; such an analysis goes beyond a static analysis of the costs and performance of a specific innovation. It needs a comprehensive analysis of the regime and of the feasibility of altering regimes. The first type of analysis consists of mapping the roles, outlook and basic assumptions of regime actors; the second type of analysis involves the identification of alternative futures and visions that are attractive from the perspective of climate protection and related societal points of view. Actors in the existing regime may have long-term visions that are interesting from a climate-protection perspective. Within the car industry, for instance, there is a converging view that in the future people will be using vehicles that run on alternative fuels, and that car use will probably no longer imply individual car ownership but instead car rental. Bill Ford, chairman of Ford, the world's second-largest car maker, has said that fuel cells will end the 100-year reign of the internal combustion engine and that we may witness an end to car ownership as the preferred method of personal transportation. In this vision, Ford and other car makers could own vehicles and make them available to fee-paying motorists when they need access to transportation (*Wall Street Journal Europe*, 6 October 2000). Government policies for climate protection could build on such ideas, because the emergence of this kind of system innovation will need governmental support. In the example given, the innovation relates not only to new technology but also to structural changes in the car-based mobility system. One of the major challenges would be to develop a car-rental system that meets the same standards of flexibility and quality as private ownership of cars.

The actual design of government policies for climate protection is no simple matter. Policies may be disruptive, or even unnecessary. An example might illustrate this. If government wants to stimulate energy-efficiency improvements in the paper industry it should take into account the concentration of R&D capacity in the relevant machine-supply industry. There are only two major European companies that supply machinery for the paper industry. The two companies almost totally control the core of the existing production technology for paper production. These industries can therefore contribute to energy-efficiency improvements from the 'inside out'. One should also bear in mind that because of their reputation and credibility, these two well-established actors are best positioned to push the traditionally conservative paper manufacturers towards innovation. Governments that want to encourage technological change in the core of the components used in paper manufacturing have to address these suppliers. These actors should not necessarily be stimulated with R&D support since a number of innovative technologies currently being developed have other (more important) advantages than energy saving. The risk of free-rider behaviour (getting money for doing things they would have done anyway) is considerable. Furthermore, it is even questionable whether they are interested in government support to fund the core of their R&D activities. An alternative might be a sort of voluntary agreement on technology development at European level. In this way, governments can try to establish energy efficiency as a higher priority on these firms' research agendas. This is just an example of how government can fine-tune intervention strategies in the existing regime. A more detailed look at the current regime of paper production also makes it clear that the slow rate of technology component replacement at a paper plant is ultimately a major obstacle to exploiting the energy-efficiency potential of 'innovative' technologies.

For these and other kinds of intervention strategies there is already a rich variety of policy instruments. Table 4.1 presents a possible classification of governmental intervention strategies and instruments.

Table 4.1 *Policy approaches and strategies*

	Supporting	**Actor interaction + learning**	**Forcing**
	Technology variation	*Coupling of variation and selection*	*Market selection*
generic	- Subsidies for R&D	- Education policy - Innovation centres	- CO_2 taxes - tradeable permits, - covenants about energy saving
specific	- Research programmes for climate-friendly innovation - Subsidies for adoption of climate-friendly technologies - Government procurement of green innovation	- Societal discussions about sustainable transport, energy etc. - Development of strategic visions and technology road maps - Strategic niche management	- Goals for renewable energy - Fuel economy standards

For the purpose of this book, the question is of course whether the strategies and instruments presented in Table 4.1 can initiate the transitional changes needed to achieve a low-CO_2 economy. The section on long term technological change showed that technological transitions are associated with structural change at different levels – companies, production chains, consumers and governmental policies – and are linked to new ideas, beliefs and sometimes even new norms and values. Many of the elements involved in transitions, such as lifestyle changes, rising income levels and changes in beliefs and values, cannot be managed. This raises the question: can a transition towards a low-CO_2 economy really be managed? The simple answer to this question is of course 'no, transitions defy control'; a more qualified answer is that although transitions cannot strictly speaking be managed, it is possible to work towards a transition, to give leverage to processes of change, and to *modulate* ongoing dynamics.

Basically there are two ways to accomplish this. Firstly, through generic policies, for example by judiciously applying economic and/or social incentives and disincentives to make some possible paths more interesting and feasible than others and vice versa. Secondly, through technology-specific policies, for example the promotion of electric vehicles through research programmes, tax credits (investment subsidies) and the use of local experiments. Technology-specific policies interfere directly in the dynamics of technical change, rather than indirectly as generic policies do. Examples of generic and specific policies are given in Table 4.1. They are arranged into three categories that are derived from an evolutionary perspective of processes of socio-technical change: policies that stimulate technology variation, those that work through the selection process and those that couple variation and selection.

For a transition to a climate-friendly economy, both types of policy approaches are needed. Generic policies such as carbon taxes are needed, especially for the uptake of climate-friendly technologies. They are in line with the economic dictum that 'prices should speak the environmental truth'. Products should be penalized according to their economic cost to society. Prices should reflect social costs. Positive externalities should be rewarded. This is an argument for the use of subsidies for technologies whose social benefits exceed the benefits for the innovator, as is often the case with climate-friendly technologies whose environmental benefits have no value in the market place.

Generic policies that change the economic frame conditions for technical change are important for transition processes towards a low-carbon economy. However, it is questionable whether they alone are able to promote the development of radical solutions that have not benefited from dynamic and scale and learning effects and adaptation of society (Kemp and Soete, 1992; Kemp, 1994). Radical solutions are often rejected and opposed by regime actors who favour incremental change (system optimization) because it is less disruptive. Therefore, the promotion of radical solutions and system innovation also requires technology-specific policies. So both types of policies are needed, and the pros and cons of each of them will differ from case to case.[5]

---

[5] A discussion of instruments is offered in Kemp (1997) and Kemp (2000).

Even then, we do not feel the combined approach of generic and specific policies is enough to produce the transitional changes required for the climate-neutral economy of 2050. System inertia can be strong, as our analysis in the previous sections of this chapter showed, and fossil fuels are still considered extremely significant for industrialized society and its development. What is lacking are clear notions of alternatives for the fossil-based development paths in industrialized society and clear perspectives on alternative technological routes. These kinds of future climate-neutral perspectives could provide a meaningful context for the development and implementation of short-term technology policies, either generic or specific. In combination with short-term policies they can fuel transitional change. What is needed, therefore, is a comprehensive new perspective on climate-oriented technology policies to align short-term policies to longer-term goals. Transition management provides such a comprehensive, integrative perspective.

The need to align short-term policies to a longer-term perspective: transition management.[6]

Drawing on the analysis of transitions in technology dynamics and systems analysis, scholars have suggested transition management as a new approach to interactively processing transitional changes (Rotmans, Kemp et al., 2000).[7] Transition management offers a methodology for organizing and processing transitional change while avoiding a deterministic and narrowly goal-oriented focus on the future. Transition management has been developed as an alternative for this kind of deterministic and goal-driven policy approach, an approach that in fact ignores uncertainties by setting policy goals based on or derived from deterministic, and often unrealistic, blueprints for future scenarios. The majority of these blueprints are unable to account for the complexities involved in transitional change. A basic premise of transition management, therefore, is that the processing of a transition should depart from and account for the uncertainties regarding the direction and content of future developments, as well as the complexities in terms of interdependencies, causalities and relationships between ecological, social and technical

---

Box 4.5
Characteristics of transition management

- long-term outlook as a framework for considering the short-term policy (*at least 25 years*).
- thinking in terms of more than one domain (*multi-domain*) and different actors (*multi-actor*) at different scale levels (*multi-level*).
- guiding and redirecting through learning processes (*learning by doing and doing by learning*).
- trying to bring about system innovation as well as system improvement.
- keeping open a large number of options (*wide playing field*).

---

[6] The section on transition management is based on the joint work of one of the authors of this paper, René Kemp, with Jan Rotmans, Marjolein van Asselt, Frank Geels and Geert Verbong. It draws on work done in Twente by the research group of Arie Rip and Johan Schot.
[7] Rotmans and Kemp, et al. (2000) give a more detailed overview and analysis of the core elements of transition management.

Box 4.6

Core elements of transition management

- choosing a collective transition objective.
- exploring quality images for a transition objective.
- formulating interim objectives.
- evaluating and learning during development rounds.
- creating social support.

systems. It has been suggested that transition management accounts for these complexities coherently, understands them and develops robust and sustainable solutions.

Transition management, therefore, actively fosters transitional change by means of interaction between governmental and private actors and is process- and content-oriented. The process is a device for interactively learning how a transition might appear and for learning new practices from the perspective of the transition (learning by doing and doing by learning).

The notion of transition management offers a framework for theory and action. Transition management is seen not as a task for government but as a task for society as a whole, whose problem-solving capabilities, imagination and support must be mobilized. At the same time, it is believed that for transitions offering collective benefits rather than private benefits for the individuals concerned, as in the case of a low-emission energy system, government has an important role to play in initiating, fostering and sustaining a transition. This role is two-fold: to realize certain substantive objectives such as CO_2 reduction (this is the *content* role) and to make sure that the process of variation-selection works well (this may be called the *process* role). The latter is important for securing desirable outcomes in the long term. The process role is aimed at stimulating and organizing the transition process, mobilizing the social actors concerned, creating opportunities and challenges for participants in the transition and creating boundary conditions within which the transition process can operate.

Box 4.6 lists the core elements of transition management, which can only be described briefly in the context of this chapter.[8] The transition goal is by definition multi-dimensional, and should take into account the different positions and ambitions of actors participating in the transition process. Furthermore, the transition goal should allow for more than one trajectory of development since one of the important aims of the process is to learn about future development options and not to achieve one particular long-term goal. Transition objectives therefore consist of a *basket of objectives* (see Figure 4.4) and is informed by quality images articulated by the actors taking part in the transition process.

The transitional track can be designed using the so-called corridor approach, not in terms of a fixed blueprint but in terms of acceptable societal risks. Societal risks can be formulated in terms of economic, socio-cultural and environmental risks (boundaries) and these risks provide a kind of dynamic corridor specifying the boundaries for future developments. These boundaries are themselves not

---

[8] An example of a transitional track along a corridor is that under no circumstances will society accept an electricity black-out.

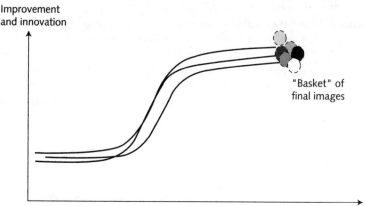

Figure 4.4 *Final objective as basket of final images*

fixed but are constantly adjusted in the context of the progress with the transition process. In this way, the corridor provides a dynamic set of frame conditions for shorter-term policy development and policies in terms of interim goals. These interim goals should not only be formulated in terms of material content, for instance, x per cent CO_2 reduction, but should also refer to achievements in the transition process itself, for instance in terms of learning effects. The interim goals should be evaluated and adjusted during so-called development rounds, when actors participating in the process analyse, assess and evaluate the learning effects regarding both process and content in the longer-term transitional perspective. Based on these evaluations, adjustments can be made regarding the transitional process, material content and the policies needed. In this way, transition management guarantees societal support, active participation by societal actors in the process and in development rounds, and the use of their knowledge and experience in the transitional process.

As already mentioned, government assumes a special position and role in the process, a role that could be labelled 'acting director'. This role combines aspects such as organizing the process, stimulating and mobilizing actor participation, process facilitation and the like. This role as director must be combined with an acting role, referring to the government's democratic responsibility for developing and implementing policies for transitional change. The different phases in the transitional process could determine the policies needed for both aspects of the governmental role as actor and director. In the pre-development phase of transitional change the governance challenge is to create a broad playing field and not just to rely on the options that are available. It is important to explore radical options. Here the goal of learning is more important than the actual greenhouse gas emission reductions achieved. Suggestions for how to do this are described in the literature on strategic niche management (Kemp et al., 1998; Weber et al., 1998; Weber and Dorda, 1999). In the take-off phase, the governance challenge concentrates on actor mobilization and motivation, for instance the interactive development of transitional goals and images. In the acceleration phase, the governance challenge is to facilitate the development and diffusion of innovations, both technical and non-technical (organizational, institutional, economic), to facilitate societal embedding of new developments and to support learning. In the stabilization phase, the governance challenge is to facilitate and support consolidation.

Learning and societal embedding are important elements of socio-technical change and hence an important aim of transition management. A thorough analysis of actors is a condition for the success of a transition process, and needs to be complemented by studies of trends and ongoing developments in niches and regimes. The various action perspectives (outlooks) of the actors should be traced and made explicit. Not only the perspectives of the existing, influential actors, but also those of the less important players who might become important in the future should be considered, as actions undertaken by niche participants may be harbingers of the future.

Our analysis shows that the aim of transition management is not so much the realization of a specific transition described in a strategic vision: it may turn out that sufficient results are obtained via the route of system improvement or that problems are not as serious as believed. The aim of transition management is to provide an active contribution to the exploration of a transition that produces collective benefits in an open, explorative manner. Transition management is not geared towards gaining control of complexity, but much more towards dealing with complexity in a productive way.

Fortunately, elements of transition management are becoming part of modern governance of socio-technical change, at least in The Netherlands. Examples are the DTO programme on innovation by means of backcasting, the COOL dialogue on climate change, which organizes and facilitates discourse among stakeholders of change, NIDO, the national initiative for sustainable development, which organizes and facilitates new initiatives (so-called jump projects) for sustainable development, and the SER proposal for a Delta programme for sustainable energy. Transition management could be a next step in this kind of learning-oriented policy approach. Recently, the fourth Dutch national environmental programme also recognized the contribution of transition management to longer-term oriented change and development for climate protection and sustainable development.

Conclusion

Drawing on notions from technology dynamics and evolutionary theory, this chapter showed that incremental (step-by-step) socio-technical change dominates patterns of socio-technical development. The climate gains of this kind of incremental socio-technical change are far from enough for the emergence of the climate-neutral economy in 2050 as sketched in both the A and B visions in Chapter 2. The climate-neutral society requires the innovation of entire systems rather than the optimization of current systems. Where system optimization dominates techno-logical-change processes, the climate-neutral society requires system innovation. The challenge for technology-oriented climate policies, therefore, is how to contribute to system innovation in a world changing and developing along patterns of system optimization. Drawing on evolutionary socio-technical change, the chapter suggested three changes.

Our first suggestion was that short-term technology policies should be based on a clear analysis of technological-change processes. Technology dynamics can help here, because a clear analysis of the dynamics of socio-technical change processes helps to develop policies that can influence (not forcefully steer) these dynamics in directions leading to system innovation. We suggested promising points of intervention for this kind of policy, being those points in technological-change processes offering potential for radical technological change. Our conclusion was that policies should focus on this potential for radical change (system innovation).

Secondly, we suggested an evolutionary-based set of policy instruments which is able to exploit the potential for system innovation. These kinds of instruments are able to initiate and facilitate the emergence of technological variation necessary for developing promising future technologies. These kinds of instruments should be combined with instruments coupling variation and selection and instruments safeguarding the market diffusion of technology by means of market selection. They can be either generic or specific in focus and content. Detailed analysis of the contingencies of technological change in specific settings should be decisive for the choice of types of instrument needed to exploit the potential for radical change in the specific setting: generic, specific or combinations of both.

Finally, we suggested transition management as a device for integrating short-term policies of technological change and innovation in the context of longer-term governance of system innovation. Transition management is a societal 'tool', offering society as a whole a methodology for organizing and facilitating transitional change towards longer-term system innovation. Transition management suggests ways of organizing and structuring the process in such a way as to ensure optimal participation by all actors involved, and it offers suggestions for how to explore and develop future images of the system innovations that society fosters. Most importantly, transition management offers a device for short-term adaptive societal learning from the perspective of longer-term system innovation. Transition management, therefore, is highly valuable for modulating dynamics of socio-technical change. By means of adaptive learning, transition management contributes to the acceleration of innovation and socio-technical change because it organizes and mobilizes the actors and capacity society needs to achieve the climate-neutral economy in 2050. Transition management is, therefore, a societal (governance) challenge rather than a governmental challenge. Government has a special role and responsibility: acting in socio-technical change and directing the process from the perspective of longer-term transitional change.

These suggestions aim to contribute to the further internalization of climate protection needs in processes of socio-technical change. Working on the internalization of these needs is in fact the important thing that can and should be done, because we don't know what the future will bring in terms of innovation and change. We only know that we want this future world much more climate-friendly than today's world. Despite our ability to describe how we like the future world to be, either A or B or perhaps C, it is impossible to order its effectuation in 2050. We can only influence and redirect the processes that will lead for certain to *a* future state of the world, instead of *the* future state of the world. The only thing we know now is that we want this future world as climate-friendly as possible and, for that reason, we can and should more actively anticipate climate protection in innovation and technological change. This chapter suggested several general ways to achieve this, knowing that it is impossible to say whether they will lead to a future world as described by the A vision or a future world as described by the B vision.

For the short term, our evolutionary-based general suggestions translate in several more specified recommendations:

- Policies can influence technological change simply because technology results from human effort, but policy intervention contains no implicit guarantee of success (emission reduction) due to the contingency of actors and factors influencing technological change. Government has little or no control over these contingencies, so centralized general regulatory approaches make no sense.

- Policies always have to intervene in existing regimes and should build on promising ongoing developments in those regimes and move from there to the further consolidation of these developments.

- Since the economic logic of the market forces actors to focus on short-term technology improvements, economic incentives for CO_2 reduction (emissions trading, CO_2 taxes) make sense because they change the economic frame conditions. However, these shorter-term policies should be supplemented with an explicit effort by public and private actors to develop an agenda for system innovation. This calls for innovative policies fostering 'learning by doing' and 'doing by learning'.

- Transitional agendas, discourse (development rounds) for evaluation and adjusting goals and means can contribute to the design of transitional processes.

- Transitions take time. Rapid and forced developments may result in societal resistance and weaken the societal legitimacy of transitional targets and means.

- Supportive policies should never back just one horse (one technological alternative). Supportive policies should build on extensive knowledge of the technological and social ins and outs of systems to learn about the potential for a climate-neutral economy and to develop policies which can support the further rise and development of this potential.

- The climate change challenge demands action now. Policies should initiate change of technology in a climate-friendly direction. Just waiting for change is no option in the time-frame of technological development.

- Technology policy should comprise both generic and specific instruments. Specific instruments, such as R&D programmes and technological experimentation, are needed for system innovation.

- Policies should not pursue one-dimensional goals, such as 80 per cent CO_2 reduction, because these kinds of goals do not reflect the wide range of changes needed for system innovation. Instead of one-dimensional goals, change processes should be guided by quality images of future developments. These images should be translated into differentiated intermediate and end goals, structuring the process towards system innovation.

References

Arentsen, M.J., V. Dinica, E. Marquart, Innovating Innovation Policy. Rethinking Green Innovation Policy in Evolutionary Perspective, *Economies et Sociétés, Serie Dynamique technologique et organisation*, no. 6, 4/2001, pp. 563-583.

Ayres, R.U. and U.E. Simonis (eds.), *Industrial Metabolism*, United Nations University Press, Tokyo, 1994.

Berkhout, F. and D. Smith, Product and the Environment: An Integrated Approach to Policy, *European Environment*, no. 9, 1999, p. 174-185.

Clayton, A., G. Spinardi and R. Williams, *Policies for Cleaner Technology. A New Agenda for Government and Industry.* European Commission, Earthscan Publications Ltd., London, 1999.

Cohen-Rosenthal, E. and T. N. McGalliard, Eco-Industrial Development: The Case of the United States, IPTS Report no. 27, Sept 1998, pp. 23-34.

Dosi, G., Technological Paradigms and Technological Trajectories: A Suggested Interpretation of the Determinants and Directions of Technical Change, *Research Policy*, no. 6, 1982, pp. 147-162.

Dosi, G., The Nature of the Innovation Process, in: G. Dosi, C. Freeman, R. Nelson, G. Silverberg and L. Soete (eds.), *Technical Change and Economic Theory*, Pinter Publishers, London, 1988.

Freeman, C. and C. Perez, Structural Crises of Adjustment, Business Cycles and Investment Behaviour, in: G. Dosi, C. Freeman, R. Nelson, G. Silverberg and L. Soete (eds.), *Technical Change and Economic Theory*, Pinter Publishers, London 1988, pp. 38-66.

Geels, F.W., Sociotechnical scenarios as a tool for reflexive technology policies: using evolutionary insights from technology studies, in: K. Sorensen and R. Williams (editors), *Concepts, spaces and tools: Recent developments in social shaping research*, Commissioned Report for the European Commission, 2000.

Geels, F. and R. Kemp, Transities vanuit sociotechnisch perspectief (Transitions from a socio-technical perspective), background document for project "Transities en transitiemanagement" (Rotmans et al, 2000), Enschede/Maastricht, The Netherlands, 2000.

Kemp, R. and L. Soete, The Greening of Technological Progress: An Evolutionary Perspective, *Futures*, no. 24(5), 1992, pp. 437-457.

Kemp, R., Technology and the Transition to Environmental Sustainability. The Problem of Technological Regime Shifts, *Futures*, no. 26(10), 1994, pp. 1023-46.

Kemp, R., J. Schot and R. Hoogma, 1998, Regime shifts to sustainability through processes of niche formation: the approach of strategic niche management, *Technology analysis and strategic management*, Vol. 10, pp. 175-196.

Kemp, R., A. Rip and J. Schot, Constructing Transition Paths Through the Management of Niches, forthcoming in: R. Garud and P. Karnoe, *Path Dependence and Creation*, Lawrence Erlbaum Associates, Mahwah, N.J., 1997.

Kemp, R., *Environmental Policy and Technical Change. A Comparison of the Technological Impact of Policy Instruments*, Edward Elgar, Cheltenham, 1997.

Kemp, R., Technology and Environmental Policy – Innovation effects of past policies and suggestions for improvement, in: *Innovation and the Environment*, OECD, Paris, 2000.

Levinthal, D.A., The slow pace of rapid technological change: Gradualism and punctuation in technological change, in: *Industrial and Corporate Change*, Vol. 7, no.2, 1998, pp. 217-247.

Luiten, E., *Beyond energy efficiency Actors, networks and government intervention in the development of energy-efficient technologies*, PhD thesis, Utrecht University, 2001.

Nelson, R.R. and S.G. Winter, In Search of Useful Theory of Innovation, *Research Policy*, no. 6, 1977, pp. 36-76.

Rip, A. and R. Kemp, Technological Change, in: S. Rayner and E.L. Malone (eds) *Human Choice and Climate Change*, Columbus, Ohio, Battelle Press, Volume 2, Ch. 6, 1998, p. 327-399.

Rotmans, J., R. Kemp, M. van Asselt, F. Geels, G. Verbong and K. Molendijk, *Transities & transitiemanagement. De casus van een emissiearme energievoorziening*, rapport voor NMP4, ICIS en MERIT, Maastricht.

Te Riele, H.R.M., S.A.M. Duifhuizen, M. Hötte, G. Zijlsta, en M.A.G. Sengers, *Transities: Kunnen drie mensen de wereld doen omslaan?*, Rapport voor Ministerie van VROM, directie milieu-strategie, Twijnstra Gudde, Amersfoort, 2000.

Utterback, J.M., *Mastering the Dynamics of Innovation*, Harvard Business School Press, Boston, Massachusetts, 1994.

Van de Poel, I., *Changing Technologies. A comparative study of eight processes of transformation of technological regimes*, Twente University Press, Enschede, PhD thesis, 1998.

Van den Ende, J. and R. Kemp, Technological transformations in history: How the computer regime grew out of existing computing regimes, *Research Policy*, Vol. 28, 1999, p. 833-851.

Weaver, P. et al., *Sustainable Technology Development*, Greenleaf Publishing, Sheffield.

Weber, Matthias, Remco Hoogma, Ben Lane and Johan Schot, *Experimenting with Sustainable Transport Innovations*, A Workbook for Strategic Niche Management, Seville/ Enschede, 1998.

Weber, M. and A. Dorda, *Strategic Niche Management: A Tool for the Market Introduction of New Transport Concepts and Technologies*, IPTS Report, Febr. 1999, pp. 20-27.

Weterings R. et al., *81 Mogelijkheden: Technologie voor duurzame ontwikkeling*, Final Report, Environmental Technology Exploration, for the Ministry of Housing, Spatial Planning and Environment (VROM) in The Netherlands, 1997.

Chapter 5

Households past and present, and opportunities for change

Henk Moll and Ans Groot-Marcus

In this chapter...

... we discuss possible implications of climate policy for households. Households are considered to be a difficult target group for environmental policy. Each individual household makes only a marginal contribution to the climate problem. The social-dilemma paradigm indicates that households are not inclined to accept individual sacrifices for the common good of a climate-neutral society. Rejecting a skeptical attitude towards the household sector (resulting in a withdrawal from addressing citizens as conscious and responsible consumers), this chapter presents an alternative view in which households can steer society in a climate-neutral direction through their consumption. With a demand-side approach (differing fundamentally from the common supply-driven approach) households can engage in processes to improve their environmental performance while maintaining their quality of life. From this perspective we will investigate the potential for, as well as the obstacles to, change in household consumption patterns and we will evaluate the opportunities to enlarge this potential and to mitigate constraints presently impeding climate-relevant changes in household consumption patterns.

Introduction

Efforts to achieve a climate-neutral society will have important implications for society as a whole. The visions A and B, discussed in Chapter 2, give impressions of the drastic changes required to attain a reduction of greenhouse gas emissions by 80 per cent. The household sector constitutes an important part of society and with its consumption contributes substantially to the use of energy and the emissions of greenhouse gases. Looking at the implications of visions A and B for households, we discern several technical measures: high level of insulation of houses, new technologies for heating, cooling and other household activities and the use of renewable energy sources. The household consumption patterns – and more generally household behaviour – are not specifically analysed and described, although visions A and B do include some general socio-cultural notions about consumption. Several projects have, however, addressed in detail the issue of reduction of energy use and of emission of greenhouse gases through changes in household consumption patterns.

Therefore, the main question we ask in this chapter is: is it possible to change household consumption patterns in a way that contributes to the attainment of a climate-neutral society? We assume that significant trend breaks are required with regard to household consumption patterns to reach the objective of a climate-neutral society in the year 2050. To achieve effective trend breaks some policy dilemmas need to be considered. The most relevant dilemmas are the multitude and diversity of actors involved (over 6 million households), the absence of the knowledge required to understand the 'global change' problem and to adopt effective measures to reduce emissions and the need for collaboration among industrial, administrative and institutional actors in developing strategies for the adoption of consumption patterns involving low greenhouse gas emissions.

To answer our question, we first discuss the present environmental performance of households in order to understand the main determinants of the greenhouse gas (GHG) emissions related to household consumption. Second, we explore the trend breaks in household consumption in the past 50 years in order to identify the main conditions determining the occurrence of trend breaks in the household sector. We then identify the options and constraints facing households in realizing radical reductions in emissions of greenhouse gases. From the results we can derive indications of the potential for achieving relevant trend breaks in the next 50 years. We will discuss four studies that have focused on possibilities for energy-saving consumption and, lastly, we will delineate opportunities for climate-relevant trend breaks.

However, the very first question is: will households survive? Von Schweitzer (1975) asked this question 25 years ago, although she posed the question for the year 2000. She assumed they would survive, and the essence of her reasoning was that households and families are more or less timeless. For centuries the basic functions of households have been to create a familiar space for everyday life and to fulfil people's daily needs. Children will always need a safe place to develop their character and skills and to socialize. Grown-ups need a place to recover, physically and mentally. There is no reason to assume that these personal and societal functions will become irrelevant in the next 50 years.

Because households are units of interacting and interdependent personalities, who have a common theme and goals, who have a commitment over time and who share resources and living space (Hook and Paolucci, 1970; Aoberg, 2000), they form a specific context for decisions and activities in everyday life and consumption. Household consumption is therefore not simply a sum of individual consumption patterns. With their common resources like money and time, skills and knowledge, their appliances and other goods, as well as facilities and services from outside organizations, they produce an output: the level of living. Households try to bring their level of living in line with the standard of living. This is a set of norms and values about daily life, partly formed by values and norms in the outside world and partly by those of the household group, by which members formulate their goals, views and ideas about the way they live. The discrepancy between the standard of living and the level of living works as a controlling mechanism. When the level of living is not up to 'standard', household production will increase, another mix of resources will be used. If that is not possible, if the resources are insufficient to maintain the 'standard', due to a decrease in salary or capacity to work, for instance, households will adapt their standards. Money plays an important role, but is only one of the resources. Human resources are equally important in reaching the household goals.

Household consumption takes place in this system of everyday life. It is directed to the well-being of the household group but also of the individual members. Individuals are only part of the system. As Deacon and Firebaugh (1988) point out, there are individual sub-systems within the household system. These systems are flexible. As children grow older they acquire more power and contribute more.

In the two visions of society in 2050 presented in Chapter 2, assumptions were made about future human behaviour. It is interesting to relate these assumptions to the household setting. In vision A people work hard, have a lot of freedom, seek self-fulfilment, are individualistic, have a progressive outlook and look for satisfaction in consumption. The market assumes parts of the household production, the activities for daily care. More meals are eaten in restaurants, there are more hairdressers and private child care centres.

In vision B the working week is less than 32 hours. People assign high values to nature, to social aspects and to justice in society; they do not seek individualistic self-fulfilment but strive for the development of a sustainable and social society, including prudent consumption, extensive public transport and collective arrangements for care of the elderly, the disabled and children.

To quote Von Schweitzer (1975) again, there are two dimensions to the way in which different households perform their functions. First, the economic-technical dimension of self-support ranges from a highly automated household at one extreme to a household that outsources as much as possible at the other. Second, the socio-cultural dimension varies from individually determined needs to socially integrated mass consumption. We can position the two visions along these two perpendicular dimensions.

Vision A seems to represent the case of fulfilment of individually determined needs combined with an emphasis on outsourcing. Vision A creates ideal conditions for young, healthy, skilled people who are able to look after themselves, have a high income and can buy all they need. They have opportunities to spend time and money on holidays and see the world. It is clear that this situation is less adapted to people with low incomes, people with physical or mental disabilities and to bringing up children.

Vision B represents the opposite end of the spectrum: self-support combined with socially integrated consumption. This vision envisages people adapted to social standards, but active in the private sphere. The young, healthy, skilled people will exercise a lot in gyms and do volunteer work in the community. Everyone who wants to work can find a place for their children in the crèche or kindergarten at a reasonable price. When children, elderly parents or neighbours are ill, it is possible to adapt work in a flexible way.

In both visions, household work will be easier than in the past. Appliances are user-friendly, materials in the house and clothing are of a high quality and easy to clean. With the market and privatization predominant in A, less time will be needed for household activities. Outsourcing becomes important but implies spending more money. Under B more time will be dedicated to household affairs and less money will be spent than under A. There is time for social activities and some household functions are performed collectively, but there is less emphasis on consumption of products that are now regarded as basic needs, such as daily meat consumption.

In a society according to vision A, life within the household seems to play a limited role in society. If the general role of households is severely reduced, it becomes questionable whether society will survive without the production of some essential functions (e.g. safety and security, education and socialization) presently delivered by the households creating at home a safe and stimulating place for the family members. Vision B is more idealistic than A, because it assumes the existence of values such as social responsibility and places less emphasis on individual material consumption, so the adherence to these values may offer better opportunities to attain a climate-neutral society. To arrive at the conditions of vision B starting from the situation as it exists in 2000, norms and values in society will have to change substantially. Norms emphasizing a care and responsibility oriented society rather than a market-oriented society need to be developed.

Direct and indirect energy consumption by households

In the Dutch energy statistics the energy consumption by households constitutes only a minor part (20–25 per cent) of the total energy consumption in The Netherlands. These household energy consumption figures only reflect the use of electricity and natural gas by households. The figures do not include energy use for private and public transport and the energy use required for the production and distribution of goods and services acquired by households.

In our analysis we adopt the central assumption that all production takes place to serve consumption. So the energy consumption and greenhouse gas emissions of the various production sectors are attributed to the households in relation to the volume of goods and services consumed by Dutch households. With this method, we follow the energy use and emissions during the complete chain of Dutch household consumption: production, transport, trade and sale, use in households and disposal.

We distinguish two flows of energy and two corresponding sources of greenhouse gas emissions. The first is the direct energy requirement, which is the energy used by households in the form of energy carriers like natural gas, electricity and petrol. The second is the indirect energy requirement, which is the amount of energy used in the production and distribution of goods and services that can be attributed to households. The total energy requirement of households is the sum of the two.

Using this approach, the average household spending in The Netherlands was analysed and the average household energy requirement was calculated for the year 1990 (Biesiot and Moll, 1995).

Figure 5.1 breaks down total household spending into different categories of spending in monetary terms. Important spending categories in monetary terms are the house, food, education and recreation and medical care. The spending on direct energy requirements is very low.

Figure 5.2 shows the energy requirements related to these spending categories. In energy terms, the share of direct energy (petrol, electricity and heating) is substantial (46 per cent). The indirect energy requirements are found in the other spending categories (54 per cent). Food, education and recreation and

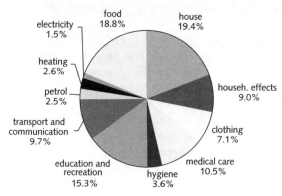

Figure 5.1 *Distribution of spending over the main spending categories of households*

household effects are the most important indirect energy categories.

The analysis is based on total household spending amounting to NLG 40,000 in 1990 (18,000 euro). The total energy requirement per household is 240 GJ/a. Around 60 per cent of The Netherlands' energy consumption is attributed in this way to households. Besides households, other relevant final energy consumers are the government and export sectors.

Additional statistical analyses demonstrated a strong correlation between calculated total energy requirement and the amount of household spending (elasticity coefficient of 0.80). They showed that there were large differences between households with an equal amount of household spending with respect to their total energy requirement (differences of up to 25 per cent from the average). Large differences also existed with regard to the energy requirements for consumption categories. For instance, the indirect energy requirement for food is substantial in relation to its share in total spending.

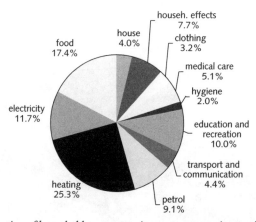

Figure 5.2 *Distribution of household energy requirements over main spending categories*

Trends in the total household energy requirement

The second half of the 20th century is very important for an analysis of present developments in household energy requirements. Economic growth and greater prosperity for households and families followed the period of post-war reconstruction in the 1950s. This resulted in a higher standard of living, better education and a desire for more possessions. The majority of households came to have more disposable income than they needed for the primary necessities of life. At the same time, family arrangements were more complex and life in the private sphere became more intensive. To ease the accompanying increase in household work, part of the money was spent on household appliances (Visser, 1969). The direct and the indirect energy requirements of households increased in line with these developments.

Van der Wal and Noorman (1998) have described the development of the household energy requirement for the period 1950–1995. Before then, during the 1940s, Dutch households had to cope with a society where, due to the Second World War, technology developments for the domestic field were at a stand-still and energy consumption was far from usual. Figure 5.3 presents the aggregated development of the direct and indirect energy requirements of households. It shows rapid growth of the direct and indirect energy requirement in the 1960s, a peak in 1979, and some decline afterwards. Both the direct and indirect energy requirement per household roughly doubled in the 1950–1995 period, notwithstanding the 40 per cent decline in the average size of households in that period. For the whole period 1950–1995 we observe that the share of direct energy in the total household energy requirement was below 50 per cent.

The direct household energy requirement is shown in more detail in Figure 5.4. Substantial increases in demand for heat, electricity and motor fuel are observed in this period. After the 1979 direct energy consumption peak, we observe a remarkable decline of the direct energy requirement mainly caused by the substantial decline in demand for heat in the period 1980–1995.

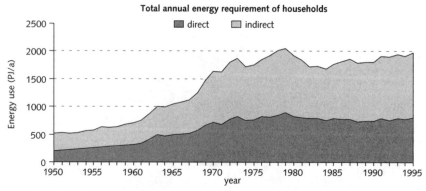

Figure 5.3 *Developments in direct and indirect energy requirements of households, 1950–1995*

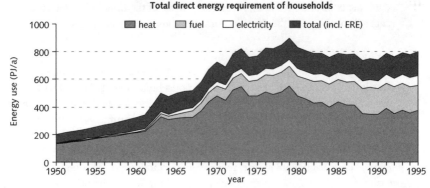

Figure 5.4 *Total direct energy requirements of households for heating purposes, electricity and motor fuel, including the energy requirement (ERE) to produce these energy carriers, 1950–1995*

To take a closer look at these trends from a micro perspective, the use of natural gas and electricity by households is considered separately. Figure 5.5 shows very steep rises from 1960 until 1980 for both energy sources. After 1980, their paths diverge: after a period of near stability between 1980 and 1990, electricity consumption resumes its steady rise. Gas consumption fell substantially during the same period, before starting to rise again after 1990.

Figure 5.5 suggests that the trends for electricity and gas consumption have different backgrounds. Electricity consumption in The Netherlands has been strongly influenced by the use of household appliances. The trend in the use of appliances was in turn constrained more by housing conditions than by availability of the appliances. Gas consumption increased tremendously after the discovery of a large natural gas field in the country during the early 1960s (see also Chapter 1 for a description of the introduction of natural gas in The Netherlands). The peak in consumption was followed by a period of energy saving, mainly due to improved insulation of houses. Both consumption patterns turned upwards again due to the

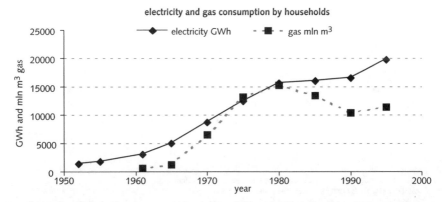

Figure 5.5 *Electricity and gas consumption in Dutch households 1950–1995 CBS, several years (109 m³ natural gas=31.7 PJ, Van Gool et al., 1987)*

economic boom of the 1990s. A careful analysis of the underlying determinants is required to recognize the most important trends and trend breaks.

Electricity

Electricity serves a wide range of household processes, mainly through the use of electrical appliances. Before 1960 Dutch households were equipped with at most one or two domestic appliances. Table 5.1 shows the development in the percentage of households with appliances – with a substantial annual electricity use – during the approximate period 1960–2000.

The washing machine was the most common appliance during the 1960s, before being overtaken by the refrigerator around 1970. The popularity of the washing machine is not surprising since washing textiles by hand is a heavy job and such tasks are the first to be mechanized in households. Similar figures, only much earlier, are seen in the USA, where the washing machine is one of the first appliances bought by households. But there it was overtaken by the refrigerator around 1940. By 1950 four out of five American families possessed a refrigerator (Lomborg, 2001). The dish washer has slowly found its place in the home during the last 25 years, while the tumble drier became widely accepted in the 1990s. This lag in ownership of dish washers in Dutch households in relation to the tumble drier contrasts with developments in most other EU countries. Only the United Kingdom and Belgium show more or less comparable figures. In France, Germany and Austria, nearly 40 per cent of households owned dish washers by the middle of 1990 and the tumble drier lagged behind in those countries (Siderius et al., 1995).

The penetration of these appliances in the households cannot solely explain the steep and steady rise in electricity use for so many years. To understand other possible factors behind it, one needs to look at how life was organized in the beginning of that period. Aspects of the daily life of housewives were investigated in 1964 in an extensive time budget study. Two facets are taken from that study: laundering and how facilities in the home limited the uptake of appliances.

As Table 5.1 shows, 75 per cent of households already had a washing machine in 1964. Of the remaining households which did not have a washing machine, most made use of laundering services or laundromats for some or all of their laun-

Table 5.1 *Electrical appliances in households from 1961 (%)*

Appliance	1961[1]	1964[2]	1972[3]	1975[3]	1980[3]	1985[3]	1990[4]	1995[4]	1997[4]
Washing machine	67	75	86	86	89	88	88	92	93
Refrigerator	12	40	88	93	98	98	97	5)	5)
Freezer	-	-	17	30	46	48	53	61	65
Dish washer	-	-	6	8	11	8	10	20	26
Tumble drier	-	-	-	-	10	14	25	41	46

[1] Margriet (1962) Lezerskringonderzoek, *Algemene analyse van de lezerskring*. De Geïllustreerde Pers N.V. Amsterdam.
[2] Philips (1966) *De Nederlandse huisvrouw*. Philips Nederland N.V., Eindhoven.
[3] Statistics Netherlands (1992) *Statistical Yearbook of The Netherlands 1992*. CBS Publications, the Hague, Netherlands
[4] Statistics Netherlands (2000) *Statistical Yearbook of The Netherlands 2000*. CBS Publications, the Hague, Netherlands
[5] Nearly all households

dry. But doing the laundry was far different from the washing we know today (see Box 5.1). In 1964, nearly half of the washing machines were not connected to a tap and/or water outlet.

At the same time, synthetic fibres and the many types of non-crease textiles were appearing on the clothing and household textile market. They did not need intensive washing, but textiles from those materials had to be washed more frequently. Twenty years later, in 1984, almost every household had an (automatic) washing machine and only 4 per cent of them had to be moved from a room, corridor or garage to somewhere else to be used. By then, about 93 per cent of the machines were connected to a drain pipe and the other 7 per cent disposed of their water via a bath or shower room. Washing clothes was much more similar to how it is done in 2000.

Clean textiles had ceased to be a major headache (Philips, 1985). This affected quantities. The amounts washed increased from just over one kilogram of laundry per capita per week in the 1960s to over five kilograms of laundry per week at the beginning of the 1980s. And it did not stop there. The volume of washing per capita per week is presently well over 6 kilograms.

So we can identify the following factors determining the huge increase of energy use for washing: less washing by hand and a growing volume of laundry per person. Explanatory determinants for these factors are behavioural change related to changes in the types of textile in use and washing frequency, technological change in the washing process (automation and electrification of heating) and increased availability of these washing technologies (affordable prices and domestic infrastructure to connect and easily use these appliances).

After the Second World War, people generally developed a demand for mechanization. The electricity system was also ready to run the appliances. By 1956, the electricity companies had connected 90 per cent of the Dutch dwelling units to the grid. In the USA, domestic appliances were already being produced in large quantities. But appliances produced in the USA were very expensive by European standards and were difficult to acquire for Dutch households. The mechanization of Dutch households only became feasible when European manufacturers started making relatively cheap domestic appliances in larger volumes.

In The Netherlands, the possession of household appliances was also influenced by another important phenomenon. Albert Heijn, one of the major grocery chains, wanted to expand its market by selling cooled food products, like the grocers in the USA. However, with such a low penetration of refrigerators in The Netherlands, people could not properly store the cooled products. Therefore in 1962 Albert Heijn started a very successful trading-stamp scheme which allowed households to save stamps to buy refrigerators and other domestic appliances at a discount. The scheme lasted for more than 10 years, and when demand for domestic appliances was saturated, the campaign shifted to appliances for do-it-yourself activities and personal care.

However, as several time budget studies of households have shown, introduction of technology may lead to time saving in individual processes, but it does not do so for whole household tasks. Although mechanization did not help to save time, it provided ease of work, comfort and opportunities to answer the greater demands of a more complex living standard. Households were better able to enjoy

Box 5.1
Households in the last decennia

In 1964, 75 per cent of households had a washing machine. But the machine of those days was markedly different from the present automatic machine. While the technology had taken over the heavy part of the washing, the actual washing and wringing, it did not finish the whole process. Housewives still had to rinse by hand.

Worse was that most houses were not designed for washing machines. Only 38 per cent of the machines were stored and used in the bathroom or scullery, and 11 per cent of the machines had to be moved every time they were used, from a bedroom, corridor or some other place to the kitchen or garage. Some people even used the machine in the open air, in the garden or on the balcony (8 per cent). And although nearly every house had a tap-water connection and a water outlet, these facilities were often not available throughout the house and could not be connected to the washing machine. Consequently, large quantities of water had to be carried to and fro.

Many housing specifications are determined by governmental regulations, which often impose minimum dimensions and facilities. In 1951 the regulations stated that every new house had to be built with a water closet, a bathroom and a wash basin. Houses are small. Towards the mid-1960s the living area was about 20 per cent larger than in 1950. During the same period the number of houses with shower or bathing facilities had grown from about 40 per cent at the beginning of the 1950s to 63 per cent in 1964. The majority of bathrooms had hot water. As regards showering or bathing, the Dutch population can be divided into two groups: one group takes a bath almost every day, the other group once a week, on Saturday. The average Dutch woman takes a bath or shower twice a week, Dutch men 1.75 times a week.

The electrical capacity in the houses is small. In 60 per cent of the houses there is not more than 20 Ampere available and that is not sufficient. About 30 per cent of people interviewed had suffered a blown fuse during the last three months. There are not enough sockets. More than half of the houses in 1964 had only one socket in the kitchen. The average kitchen had 1.6 sockets. People mention the lack of sockets as a problem.

During the 1950s Dutch living rooms were heated by coal stoves. Coal then had to be carried in every day. The stoves constantly produced dust. Only about 5 per cent of the houses had central heating, also coal-fired, which meant a lot of work. By the end of the 1950s, people gradually started to use oil. Oil is easier to handle, but more expensive. This development was interrupted by the discovery of large amounts of natural gas on Dutch territory in 1962. At that time, in about 80 per cent of the houses the living room was still heated by a coal stove. About three-quarters of the families had a second heater that ran on electricity or paraffin to warm a second room. These were only occasionally used. Normally families stayed together in the living room in the evening (87 per cent), usually watching the television together if they had a television set (64 per cent).

Philips (1966) *De Nederlandse huisvrouw.* Philips Nederland N.V., Eindhoven.

a higher standard of living (Groot-Marcus, 1984). Technology improves the quality of life of the person who performs household tasks and of the other members of the household at the same time.

Because of constraints in the domestic infrastructure and in the supply of appliances, household consumption of electricity did not explode but increased gradually in the second half of the 20th century. The facilities for washing became more and more sophisticated, as did the domestic infrastructure for showering and bathing.

What is important in the context of washing clothes – and also with regard to showering and bathing – are the values and norms for hygiene and cleanliness. Quality standards in households are not fixed. When a certain level of quality is reached and only a moderate effort has been needed to get there because of technological developments, the goals are set anew. There seems to be no limit to growth. This is apparent in the cases of bathing and washing. It is not the necessity of hygiene that dictates the frequency of bathing, showering and washing, but the change in norms and the ease with which they can be met.

Housing

Because of the housing shortage after the Second World War, which lasted well into the 1960s, houses were too small and inflexible to accommodate an expanding number of appliances per family (Visser, 1969). Around 1965, the minimum requirements for new houses were raised, and towards the end of the 1980s the minimum floor space was increased again. Broadly speaking, the houses of the 1990s are about 40 per cent larger than 40 years earlier, while the number of persons per household has declined at the same pace (from nearly 4 persons per dwelling to 2.45 in 1994). Today, nearly all houses have a shower or bathroom, and sometimes more than one, and they are better equipped. For one room the government did not raise the minimum dimensions: the kitchen. In 1965 kitchens could be built with just 5 m² of floor space, and that was still the case in 1990. Kitchens with the minimum dimensions are designed for two appliances, an oven and a refrigerator (van der Meer, 1989). Scarcity of space may explain the slow acceptance of dish washers in Dutch homes and the preference for buying a tumble drier, which can be placed elsewhere.

Heating

Heating has a rather different history than electricity use. Dutch houses switched almost completely from heating with coal stoves to central heating with natural gas in less than 20 years. This changed life quite dramatically in the 1960s (see Box 5.1). The availability of cheap natural gas led to government campaigns in the 1960 to persuade households to use gas for cooking and heating. Central heating was also promoted, resulting in 30 per cent of houses being centrally heated by 1971. By 1984, 75 per cent of houses were centrally heated by natural gas. Most of the 13 per cent of houses that were not centrally heated in 1995 used gas stoves in more than one room.

After the massive switch to central heating, the consumption of gas became exorbitantly high (see Figure 5.3). At the peak of the domestic gas consumption, around 1980, the required energy per household went up to 70 GJ/a. Stimulated by an extensive publicity campaign and government subsidies, in the 1980s people started to insulate their houses. In the 1990s, annual energy use declined to about 40 GJ per 2.5-person household. This amounts to 16-18 GJ per person, which, given that the total energy requirement of a four-person household for heating, cooking and lighting was about 25 GJ a year in the middle of the 1960s, was three times more than 35 years earlier, a substantial rise in energy consumption

We can identify the following factors that determine the increase in gas consumption by households. The easy availability of natural gas triggered a lot of behavioural changes including a longer heating season, an increase in the average indoor temperature, a substantial rise in the number of rooms heated in the house and an increase in the use of heated water for bathing and showering. The expansion of thermal insulation in houses reversed - especially in the period of high energy prices - the rising trend of natural gas consumption.

We can also see that central heating created opportunities to exploit much more living space. This development cannot be reversed. Dutch households cannot live their lives in one room any more. Current social norms dictate that every member of a household has the right to personal development and privacy.

Trends and trend breaks with regard to direct energy consumption in the 1960-1995 period

Consumer behaviour is determined by three main factors: individual and household needs, opportunities (external facilitating conditions) and abilities (internal capacities) (see Gatersleben and Vlek, 1998 for a description of this explanatory model).

The trends in gas and electricity consumption are related to changes in these factors. First, trends relating to 'opportunities' (i.e., the connection of households to the natural-gas pipeline system and to the electricity grid and the availability of household appliances and heating systems) partly explain the substantial increase in the direct household energy use. But the increase in the 'abilities' of households (related to the rise of personal income and the increase of interior space per household member) is also important in explaining the rise of direct household energy use. Another interesting factor is the observed feedback from the 'opportunities' and 'abilities' factors to the 'needs' factor. The needs also changed during this period: a need for privacy related to the increased availability of (heated) space, a need for higher standards of hygiene and cleanliness because of the opportunities to achieve such higher standards. So with regard to the direct energy use in households we observe a co-evolutionary process: economic changes, changes in infrastructure and socio-cultural changes (affected by the changes in individual and household needs) influence each other in an intricate way, resulting in the steeply rising trend of direct energy consumption in the 1960–1980 period. Some (temporary) trend breaks after 1980 are also explained by these factors. The economic recession during the 1980s temporarily limited the growth of electricity use (through decreased financial abilities). The availability of newly developed energy-saving technologies (insulation, highly efficient heating appliances) curbed the use of natural gas.

These and other social trends determine the elements of the direct and indirect household energy requirement not discussed so far. The greater use of motor fuels can be analysed in terms of a similar co-evolutionary perspective. A closer inspection of the determinants of the changes with regard to the indirect energy requirements yields interesting insights: relative increases in the categories 'leisure and education', 'personal care' and 'household furniture and decoration'. These findings are related to social trends like individualization, emancipation, emphasis on personal development and increase of leisure time.

Notwithstanding a general increase in environmental awareness in society, its influence on direct and indirect household energy consumption appears to be marginal. The energy use for heating may have fundamentally declined in part for environmental reasons, but although the use of electricity, motor fuel and the indirect energy use declined for some years because of an economic recession, they resumed a steady growth path in the succeeding period of economic recovery.

Feasibility of substantial greenhouse gas emission reductions by households?

First, this chapter presented an accounting for the total household energy requirement. These accounts demonstrated that a substantial portion of Dutch energy consumption should be allocated to households. The previous section described and discussed some trends that led to a substantial increase in household energy requirements in the decades after the Second World War. Substantial reductions are also required within households to meet the long-term emission targets for greenhouse gases. Based on this notion, several research programmes have been carried out to assess the potential for reduction of energy use and/or greenhouse gas emissions through changes in households. We will discuss four of these studies: the Perspective project, the GreenHouse programme, the HOMES programme and the Sushouse programme. These studies adopted an integrated approach, addressing the (macro-level) relationships between sectors and the relationship between production and consumption.

Several investigations identify options to reduce the household energy requirements and the related greenhouse gas emissions using the distinction between direct and indirect energy requirement.

The *direct* energy requirements of households can be reduced along two lines. The first line, a technological line, is directed at technological improvement of household appliances (low-energy refrigerators, high-efficiency heaters and boilers), improvement of the physical infrastructure (increasing efficiency in electricity production, reducing the need for physical transport), and investments to save on fossil energy (insulation of dwellings, use of waste-heat sources, use of renewable energy). The second line, a behavioural line, is directed at bringing about changes in household behaviour leading to reduced use of appliances and lighting, lower temperatures in space heating and water heating (for clothes washing), and in possession of fewer appliances.

The *indirect* energy requirements of households can be reduced along the same lines. The technological line is targeted at improving energy efficiency and saving fossil energy in the production and trade sectors; the behavioural line is aimed at changing the consumption pattern into a low energy-intensity pattern, such as buying green products, buying products with a low energy intensity and substitution between spending categories.

Perspective project

The 'Perspective' project involved a practical study of the possibility of reducing energy consumption through information-induced behavioural change of consumption patterns (see Schmidt and Postma, 1999). Twelve households participated in a two-year field experiment. Their target was to reduce direct and indirect energy consumption by 40 per cent, notwithstanding the 20 per cent rise in income they were awarded for meeting the reduction target. Households had to fill in weekly lists concerning their energy use and shopping and received direct feedback through informative software on the energy consequences of their behaviour. In addition, tutors gave intensive counselling to the households during the experiment.

The households almost succeeded in meeting the 40 per cent reduction target. Most of the reduction related to the indirect energy requirement. The direct energy requirement was reduced by energy-saving behaviour and by replacing old appliances with more energy-efficient ones. Important changes observed in consumption were:

- increased purchase of expensive labour-intensive goods (designer goods, handmade goods and works of art);
- increased purchase of high-quality and durable goods (clothes, shoes and furniture) and extension of the lifetime by repair;
- alteration of diets (less meat, biologically produced food, and seasonal vegetables instead of vegetables produced in greenhouses);
- changes in the mobility pattern, including mobility reduction (short-distance holiday locations and cycling for commuting);
- increased purchase of services (domestic help, visiting restaurants and education, beauty services).

Lifestyle changes arising from this experiment were generally regarded as positive. The purchase of high-quality goods and personal services formed a very attractive ingredient of the new lifestyle. The participants initially found the constant attention to energy while shopping to be a burden, but became accustomed to this aspect of the experiment later. The attitude to the restriction of mobility was negative throughout the experimental period.

The lifestyle of the participants was also analysed one and half years after the period of the experiment in a follow-up survey. Changes in diet had been maintained, partly to almost entirely. Changes with respect to purchases of high-quality goods and personal services were partially maintained. Most of the changes regarding transportation and holidays had been abandoned. In addition, the direct energy consumption had increased moderately. Two setbacks can explain these results. The discontinuation of the financial contribution limited the possibility of spending money on quality goods and services. The discontinuation of control and feedback eased the pressure to maintain the discipline required for adherence to an energy-saving lifestyle. The participants gladly switched back to an intensive mobility pattern after the experimental period.

GreenHouse programme

The GreenHouse programme (see Moll, 2000; Uitdenbogerd et al., 1998; Nonhebel and Moll, 2001) was designed to reach an integral assessment of the medium-term potential for reducing the energy requirement and greenhouse gas emissions of households. This programme defined a comprehensive package of reduction options that are potentially feasible in the period 2000–2020. With a constant average household budget, the potential reduction is 54 per cent in energy terms and about 50 per cent in terms of greenhouse gas emissions. Given the assumed increase in incomes (2.7 per cent annual growth rate of GDP) and energy efficiency (1.7 per cent annual decrease in energy intensity) according to the European co-ordination scenario of the CPB, the absolute reductions achieved with the 'GreenHouse package' compared to the 1995 level are about 10 per cent for energy and greenhouse gas emissions in 2020 (see Wilting et al., 1999).

The package consists of technical improvements aimed at reduction of direct energy use (20 per cent), behavioural measures aimed at reduction of direct energy use (3 per cent), technical improvements in the production and distribution of goods and services aimed at reduction of indirect energy use (26 per cent) and changes in the consumption pattern effecting changes in the indirect energy requirement (5 per cent). About half of the reduction could be achieved by measures taken in the production sectors (supply-side options); the other half should be achieved by changes within the households (demand-side options).

This package is largely based on technical measures, and the role of behavioural change seems negligible. Of course, some technical measures also have a behavioural component: the acquisition of an energy-efficient refrigerator is also an aspect of behaviour, depending on the price, the availability, information and so on.

Complementary to the package design, a survey was conducted among households to investigate their knowledge about and the feasibility and acceptability of the various reduction options. According to the survey, the most well-known options were energy-saving light bulbs, driving less in a car, using a bicycle, lowering average room temperature, water-saving shower heads and washing at lower temperatures. The set of less-well-known options included replacing frozen or fresh imported vegetables with preserved vegetables.

The acceptability of the reduction options in the survey was measured in two ways. For each option, the respondents were first asked to give an acceptance percentage ranging from 0 to 100 per cent. At the end of the survey the respondents were then asked to choose four reduction options out of 32 that they would like to try and four options they were most reluctant about. Because the survey provided many details regarding the consequences of the options, respondents could estimate the implications for their own situation quite well.

The most popular reduction options were energy-saving light bulbs, energy-efficient household appliances, replacement of cut flowers with plants or other presents, water-saving shower heads, replacement of greenhouse/imported vegetables with preserved vegetables. The less popular options were no car, no freezer, no drier and greater use of public transport.

Although options were designed in such a way that implementation would have a limited effect on household practices, the willingness of households to implement the suggested options was very limited. Only options with barely any

effect on household practices (such as the use of energy-saving light bulbs, purchase of an energy-efficient washing machine etc.) could be interpreted as feasible
under present conditions. There was one exception, involving the composition of
the menu. Changing to less greenhouse gas-intensive vegetables (seasonal products) seemed a realistic option. Also striking was the fact that none of the options
that affected mobility proved feasible.

The survey results demonstrate that, without any additional measures in
infrastructure etc., on average only a part of the demand-side potential for reduction of greenhouse gas emissions is currently feasible: an estimated 5 per cent–10
per cent of total household energy requirements. This estimate is much lower than
the indication (around 25–30 per cent) based on technical analysis of the envisaged
reduction package (Uitdenbogerd et al., 1998).

HOMES programme

The HOMES programme (Noorman and Schoot Uiterkamp, 1998) addressed the
issue of sustainable household metabolism. This issue was elaborated in some
aspects of the household consumption pattern: space and water heating, the acquisition and use of (electrical) appliances and transport. A multidisciplinary perspective was followed, including socio-psychological, spatial planning, policy and
energy/environmental analysis. From each of the disciplinary perspectives, the
programme presented a vision of the possibilities and constraints with regard to
realizing a more sustainable household consumption pattern. Some important
findings are presented below.

The energy analysis focused mainly on direct household energy use. A scenario approach demonstrated a potential 40 per cent reduction of direct household
energy requirements in the Towards Sustainability scenario, and a 10 per cent
increase of direct household energy requirement in the Business-As-Usual scenario (Van der Wal and Moll, 2001).

On the basis of a survey, the socio-psychological analysis demonstrated that
individual respondents may accept moderate reductions (10–15 per cent) of their
total energy requirements in order to achieve a more sustainable society.
Respondents also believed that reductions of up to approximately 25 per cent will
not have a major effect on their quality of life (Gatersleben, 2001).

The spatial planning analysis addressed the relationship between forms of
urbanization and household energy requirements. It emerged that the position of
the urban centre in its surroundings is an important determinant of a household's
mobility in practice. For a regional centre within a largely rural area, compact
urban development demonstrates the emergence of favourable mobility patterns.
In other areas that already have high-density urban centres (the so-called urban
fields), new compact urban areas will stimulate mobility because of the large diversity of spatial relations in that case (Van Diepen, 2000).

The policy analysis addressed the issue of steering households toward a more
sustainable consumption pattern. Here, the production-consumption chain is also
very important. All actors (producers, retailers, consumers) in that chain should be
targeted. Steering producers and other economic actors is easier than steering the
consumer because of the wide diversity of the latter group and also because of the
remoteness of the government from the consumers. Indirect mechanisms

(through economic actors) and unilateral mechanisms (taxation and information) were recommended to steer consumers effectively (Ligteringen, 1999).

SusHouse programme

The SusHouse programme analysed the long-term feasibility of sustainable household practices (for food, see Quist, 2000; Dekker, 1999). Potentially sustainable scenarios were derived for some important household activities from a mixture of forecasting and backcasting techniques. The forecasting process identified major economic and socio-cultural trends for some household activities. Based on these trends, some long-term scenarios were constructed. In the second stage, the consistency of these scenarios with sustainability goals was analysed, with the aim of assessing social and technical conditions crucial to meeting the sustainability goals within these scenarios. In the third stage, the backcasting, ways of reaching a future state corresponding with the sustainable scenarios were explored.

With regard to food, the SusHouse project produced three greatly differing scenarios: the intelligent kitchen, the 'superrant', and local food production.

1. Acceleration and intelligent high-technology cooking: Households are characterized by high-tech, convenience, do-it-yourself, moderate use of services and acceleration of the pace of life, especially during the working week. There are no fixed meal times, but meal periods during which one can have meals depending on working hours and leisure activities. Environmental gains arise from management optimization in such a system and the use of sustainable ingredients.

2. The neighbourhood food centre (also named the superrant): In a compact city people can go to centres for a meal, grocery shopping, to purchase take-away meals and eat together for different prices. This neighbourhood food centre is visited by a wide variety of households including single people, families and elderly people. Environmental benefits stem from local up-scaling, sustainably-grown foods and increasing the service component.

3. Local green menu through green consumer demand: Households buy and eat seasonal food grown locally and purchased at local shops, small supermarkets and from the grower as 'fresh' unprocessed ingredients. Imported products are expensive due to the incorporation of environmental costs. The coupling between food and origin is enhanced and all the cooking is done in the kitchen at home. Environmental benefits arise from sustainable, seasonal and locally grown foods, accompanied by the minimization of transportation and processing.

Assessing these scenarios against sustainability goals, for instance with regard to greenhouse gas emissions, with each scenario there are major difficulties in meeting these goals, and additional far-reaching assumptions are required, for instance, as regards efficiency of appliances and personal and goods transport in the first two scenarios and with regard to the personal diet in the third one.

Reviewing these four partly complementary studies, we can draw some conclusions about the reduction potential related to household consumption.

The Perspective project shows a substantial potential for behavioural change, effectively resulting in substantial reductions even in the case of growing personal income. Comparable results are found with Eco-teams, using information transfer, self-monitoring and social stimulation (see also Chapter 3). But we should be aware of the considerable initial effort put in by the households participating in the Perspective project and by Eco-team members. The induced behavioural change will also only be partially maintained without continuation of the special attention given to the people and constant flows of feedback and information.

The GreenHouse and HOMES programmes followed a mixed approach to reducing greenhouse gas emissions: technological development and implementation targeted at energy efficiency and behavioural change. In the event of growing personal income, the gains of this combination of approaches are severely reduced. Nor is there a high degree of acceptability for the assumed behavioural change for most of the reduction options. The HOMES programme demonstrates that additional approaches – beyond the technological and behavioural approaches – such as an institutional or a policy approach and a spatial planning approach may offer better prospects for far-reaching reductions. Changing the institutional and physical infrastructure around the households may accommodate more environmentally friendly and energy-saving behaviour than the present social and physical environment of households. An example is provided by the study about deprivation of the rural poor due to their inability to travel in the UK. To create sufficient possibilities for them for a decent way of everyday living, a policy package is needed consisting of an adequate range of services. First, facilities at local level need to be supported by the village community or local authorities (grocery shops, library, banks, medical services); second, better public transport, supported by national regulations as well as by local authorities; and third, compensation in travelling costs for individuals with a low income. Better public transport will also lure car owners out of their cars. The challenge here is to integrate the measures implemented by different actors (Boardman, 1999).

The scenarios of the Sushouse project assume new socio-cultural standards as well as a new infrastructure. To achieve a SusHouse scenario, trend breaks are required because of the large differences between the scenarios and present-day standards and infrastructure. The results of an evaluation of the SusHouse scenarios again demonstrate the difficulties of realizing absolute reductions in a world of growing consumption.

The overview of these programmes may also provide insight into the characteristics of the options at the household level that would yield reductions of greenhouse gas emissions and are acceptable to a substantial share of households. Technical options based on increased efficiency of appliances and light bulbs are generally acceptable. Intensification of energy-saving behavioural patterns - already existing in the household - with regard to the use of appliances is accepted in many cases. With regard to indirect energy use, many substitution options (giving other types of presents, some changes in food consumption) are also easily accepted when enough information is available. The acceptance of some other options is very low, such as the abandonment of appliances, the reduction of car use and the restriction of choice with regard to holiday destination. Characteristics of such options are that they are perceived to affect the quality of life and to limit the households' ability to meet their standard of living.

Voluntary moderation of consumption, although in principle a very effective way of reducing emissions of greenhouse gases, is not adequately addressed in these programmes. The Perspective project was designed to shift energy and material consumption to high-quality and energy-extensive services, but did not emphasize moderation of consumption per se. Some options in the GreenHouse and HOMES programmes addressed reduced possession and use of appliances but not general moderation of consumption. Voluntary moderation of consumption seems to be contrary to the present-day economic growth trends and socio-cultural norms. So, for the time being we cannot expect much success from the stimulation of voluntary moderation of consumption.

With respect to the dilemmas discussed in Chapter 3, some remarks can be made about the feasibility of substantial greenhouse gas reductions by households. Not all dilemmas are relevant at this level.

The multitude of actors in the household sector poses a severe steering problem. According to Ligteringen (1999), indirect (unilateral) strategies are required for steering in such a case. The 'centrally-directed technological' trajectory, with a small number of actors involved in decision-making and implementation, is potentially a satisfactory strategy for reaching reduction targets. However, this strategy also poses the risk of the rebound effect, spurring the growth of consumption. Additional, indirect strategies should be included, aimed at securing changes in the social and physical infrastructure that facilitate household behaviour which leads to reduction of greenhouse gas emissions.

It is also clear that by contrast with other sectors in society the household sector is characterized by a diversity of small actors with no expertise in the energy field. The diversity of ideas, goals and contexts that exist in everyday life poses severe problems for efforts to get people to commit to a coherent and broadly accepted reduction strategy. Moreover, besides the layman's lack of knowledge of ways to change behaviour, everyday life nowadays is quite complex and dependent on arrangements made in different fields of society. In order to reach the current level of living and consumption, in the course of time many supporting organizations and infrastructures have been established and institutionalized. It is very difficult for those underneath to escape from this entrenched way of living. De-institutionalization will be more feasible if the lead comes from other social levels, such as local government.

The dilemmas about creating awareness and building up support through integrated sustainable development strategies are also relevant because of the multitude of actors and because of the lack of common knowledge and understanding about the climate issue as well as the sustainable development implications. The Perspective project and the 'Eco-teams' demonstrate the possibility of resolving these dilemmas in specific settings.

The co-benefit approach mentioned in the sixth dilemma would be more successful in the case of households. The first concern of a household will always be to match its level of living to its chosen standard of living with regard to the satisfaction of everyday needs of the household members. Households have developed skills, structured their lives, formed ideas and habits and acquired resources to reach this goal. If this goal can be achieved better or more easily by measures with a co-benefit for the environment, they will happily accept them. All programmes discussed above demonstrated much higher acceptance of options with co-benefit potential.

We can conclude with the following findings. The contemporary trends in society and the economy will not generate autonomous reductions of the greenhouse gas emissions related to household consumption. The growth of personal income combined with a low acceptance of energy-saving behaviour will nullify the expected efficiency gains in production and consumption technology. On the other hand, it has been demonstrated that substantial reductions through changes in the household sector are feasible under favourable conditions.

Opportunities for climate-relevant trend breaks in household consumption

The research with regard to sustainable household consumption demonstrates that a contribution by households to a societal development in accordance with vision A or vision B is doubtful.

A development according to vision A corresponds to present-day standards in households. These standards largely determine the current trends in society. Extrapolation of these trends does not produce climate-neutral household consumption scenarios. The additional technological trend breaks in vision A – over and above common Business-as-Usual scenarios – will probably spur household consumption further because this vision will not uncouple technological development and consumption growth. In this vision the consumer is not held responsible for his/her consumption. How can we expect the market parties and the low-profile government to take full responsibility for the reductions? A more fundamental question is who will take responsibility for the reproduction of the human population if individual short-term motives are dominant and the values necessary to create and maintain family life are almost absent. And how will the social infrastructure be reproduced if the investments required are only profitable in the long term?

A development according to vision B seems to be more effective than vision A to achieve substantial long-term reductions and to facilitate climate-neutral household consumption. However, the social values, responsibility and moderation assumed in that vision are fairly inconsistent with present social values and aspirations. So a social development in accordance with this vision will lack general acceptance among the present population. Moreover, the contemporary physical, institutional and social infrastructure does not accommodate a responsible and moderate lifestyle. So radical social and institutional trend breaks are required because the changing of social values and the renovation of infrastructure both require a long time span and constant intensive efforts.

The achievement of climate-neutral household consumption patterns therefore requires trend breaks at the technological as well as the social and institutional levels. In other words, a co-evolution of parts of vision A and B. A co-evolutionary development of technical and social systems has occurred before in The Netherlands. The description of trends in household consumption demonstrated the simultaneous occurrence of far-reaching trend breaks at the technical and the social level. Households did not intentionally increase their energy consumption, because energy had a price, and in part of that period even a very high price. The increase in household energy use was mainly due to the implementation and growth of new technical systems and radical social changes, such as the construction of the natural-gas pipeline system, new standards for the building of houses, new policies on spatial and road infrastructure planning, and the general democratization of society and the emancipation of women. A deliberate design of future technical systems and institutional and social measures with the indirect effect of reducing energy use may prepare the conditions for climate-neutral consumption patterns by households. Parts of visions A and B can be used to develop this approach.

Present-day societies are plurifom, made up of groups of people with different cultures, needs, ideas and standards. Ideas and needs also change during a person's life. Nor are countries and regions uniform; there will always be regions where people live close together and more rural areas. Positive appreciation of this pluriformity should also function as an antidote to the mechanism of social comparison which has stimulated growth of consumption in the past.

Vision A focuses on an important segment in present-day society: young, active people with a reasonable income living in smaller or larger towns. Climate-neutral consumption patterns imply that these groups in particular should give great attention to energy efficiency in their consumption and to the use of renewable sources in production and consumption technology. Adaptation will not be difficult if it can be advertised as typical consumption for their peer group. The implementation of these technologies is strengthened and the risk of rebound behaviour is mitigated by the implementation of regulatory eco-taxes on producer and consumer energy consumption. Producers are allowed to include the taxes in the consumer price. In that way, the prices of direct and indirect energy consumption are equalized at the consumer level. Full energy pricing may steer consumers to more responsible behaviour regarding energy use.

Vision B mainly addresses groups of people who are more concerned for their surroundings and the networks around them. These households should be informed and stimulated to behave in a responsible way. Emphasis should be given to the dissemination of information about the need for and advantages of new technologies for energy and CO_2 reduction. The development of alternative energy sources should have broad support among these groups. Private individuals, companies and households should see the necessity of such developments and people who have money should act as innovators, buying and trying out newly developed photovoltaic systems, sun collectors or windmills. Information must be made available in a variety of ways (labels on products, individual information and calculation software, tutoring groups to change behaviour) and on multiple levels (education, community centres, work place). In addition, incentives should be created to buy energy-saving products, to experiment with energy-friendly behaviour and to maintain and improve environmental lifestyles.

To achieve climate-neutral consumption patterns it is very important to give sufficient attention to the structures in society that were formed at other times with different goals, norms and values as well as technology in the field of energy production and use. The physical, social and cultural infrastructure should be deliberately reconstructed in such a way that responsible behaviour is facilitated and undesirable behaviour is curtailed. In many cases, the co-production strategy should be worked out. Restructuring the parking and traffic infrastructure of neighbourhoods and a deliberate choice of location for social facility centres may induce changes in the modal split together with an agreeable local environment and an easing of the time-stress on households. The deliberate design of environmentally-friendly service facilities may enhance the quality of life as well as reducing energy use.

Many trend breaks are necessary to achieve climate-neutral consumption patterns. A co-evolutionary process is required to realize radical emission reductions. As we argued in this chapter, consumer behaviour is determined by needs, opportunities and abilities. Changing needs is vital for a change in behaviour, but it does not have to start there. It appears that in the last 50 years changes in needs have been triggered by developments in the opportunities and abilities of households. It must be remembered that the co-evolutionary process seen in the last century was aimed at growth rather than reduction of material consumption and energy use. But an emphasis on reduction is not appealing to people, so attention must be drawn to the benefits that will ensue. In the anterior growth path, technology took the lead and the political and social system followed. Nowadays there are signs that the technology-push is losing its influence and that technology is developed more and more in interaction with the societal and (maybe) environmental demand. In the future reduction path, government policy should take the lead because of the inability of technological and social systems to initiate processes that will achieve absolute reductions.

What strategies should be used to resolve the dilemmas and push households in the direction of a climate-neutral policy? Based on the insight of the research programmes on more sustainable consumption, we can present three policy approaches that should be adopted together.

First, make available general information about the climate system and the role of energy consumption as well as information about specific energy-saving options at technical and behavioural level. This information must reach every consumer who might be inclined to consume with climate responsibility. Various information and implementation strategies should be followed to reach interested households: by Internet and software-based tutoring directed at individuals, dedicated information for city and neighbourhood groups and the formation of local teams practising climate-neutral consumption, the promotion of energy-efficient consumption in magazines of grocery retailers, and so on. In this it is important that elements of the energy-saving consumption patterns should be associated as far as possible with other values, such as health, quality and innovativeness.

Second, national and local government should stimulate and implement indirect strategies to change household behaviour. Direct strategies are difficult and ineffective because of the enormous number of households and the pluriformity of the household sector. Some indirect strategies should address other actors, such as producers and retailers as well as institutional organizations representing segments of the household sector. Other indirect strategies could influence market conditions with a view to achieving reductions in direct and indirect energy requirements. Taxation of energy and of consumer goods and setting standards with regard to the direct and indirect energy requirements of products are potentially effective ways of steering consumption.

Third, local and national government should adapt the general conditions for household consumption in a climate-neutral direction. The physical infrastructure (houses, roads and transport systems, energy systems) should be renovated and restructured to facilitate an agreeable lifestyle and climate-neutral behaviour. The co-benefit approach is very important. Solutions for local problems with regard to quality of life should be linked to the global climate issue. The production and trade sectors should be encouraged to shift the present emphasis on possession of goods by consumers to an emphasis on the delivery of services and to take full responsibility for the product life cycle. The development of climate-neutral product-service systems may also generate co-benefits.

There are opportunities for the emergence of a climate-neutral consumption pattern by households, but in most cases more is needed than goodwill from households alone. The achievement of climate-neutral consumption presents a major challenge, for society as a whole to strive for this goal, and for national and local government to develop strategies and secure support for organizing society in an environmentally sound way.

References

Aoberg, H., *Sustainable waste management in households – from international policy to everyday practice, Experiences from Swedish field studies,* Ph. D. Thesis, Göteborg Studies in Educational Sciences 150, Göteborg, Sweden, 2000.

Biesiot, W. and H.C. Moll, *Reduction of CO_2 emissions by lifestyle changes.* IVEM onderzoeksrapport no. 80, Groningen, 1995.

Boardman, B., Rural transport policy and equity, in: *Energy Efficiency and CO_2 Reduction: The dimension of social challenge, Proceedings of the 1999 ECEEE Summer Study, Mandelieu, May 31 - June 4 1999, Paris,* Ademe Editions, European Council for Energy Economy, Part II, Panel V, 10, 1999.

Boer, J. de, P. Ester, C. Mindell, M. Schopman, *Energiebesparings programma's ten behoeve van consumenten in Nederland.* SWOKA onderzoekssrapporten, no. 16, 's Gravenhage, 1982.

Deacon, R.E. and F.M.Firebaugh, *Family resource management: principles and applications.* Boston: Allyn and Bacon, 1988.

Dekker, C.P., *Consumer acceptance of future scenarios in food-consumption, An explorative research,* WUR Household and Consumer Studies/SusHouse, Wageningen, 1999.

Diepen, A.M.L. van, *Households and their spatialenergetic practices. Searching for sustainable urban forms,* Ph. D. Thesis, University of Groningen 2000.

Gatersleben, B. and Ch. Vlek, Household consumption, quality of life and environmental impacts, in: Noorman, K.J. and T. Schoot Uiterkamp, *Green Households? Domestic consumers, environment and sustainability,* London: Earthscan Publications, 1998.

Gatersleben, B.C.M., *Sustainable household consumption and the quality of life: examining the perceived social sustainability of environmentally sustainable household consumption patterns,* Ph. D. Thesis, University of Groningen, 2001.

Groot-Marcus, J.P., Technologie, een oplossing voor de huishoudelijke arbeid?, *Wetenschap en Samenleving,* 1984, nr 2, pp 28-37.

Hook, N.C. and B. Paolucci, The family as an ecosystem. *Journal of Home Economics* 62, 1970, pp. 315-318.

Ligteringen, J.J., *The feasibility of Dutch policy instruments,* Ph. D. Thesis, Twente University Press, 1999.

Lomborg, B., *The sceptical environmentalist, Measuring the real state of the world,* Cambridge University Press, Cambridge, 2001.

Margriet, *Lezerskringonderzoek, Algemene analyse van de lezerskring.* De Geïllustreerde Pers N.V. Amsterdam, 1962.

Meer, A.van der, Garanties voor woonkwaliteit?, in: J.M. van Dam et al (eds.), *Woonecologie tussen consumptie en existentie,* SHVW, Wageningen, 1989.

Moll, H.C., Multi level evaluation of options to reduce greenhouse gas emissions by change of household consumption patterns. In: *Proceedings of the third biennial conference of the European Society for Ecological Economics: Transition towards a sustainable Europe. Vienna, May 3 to 6, 2000,* (full text on CD; abstract in proceedings book pp. 92), 2000.

Nonhebel S. and H.C. Moll, *Evaluation of options for reduction of greenhouse gas emissions by changes in household consumption patterns; Final report GreenHouse project.* NRP report 410 200 059, 2001.

Noorman, K.J. and T. Schoot Uiterkamp, *Green Households? Domestic consumers, environment and sustainability*, London: Earthscan Publications, 1998.

Philips, *De Nederlandse huisvrouw*, Philips Nederland N.V., Eindhoven, 1966.

Philips, *De Nederlandse huisvrouw 20 jaar late*, Een samenvatting ontleend aan gegevens die aan Philips ter beschikking zijn gesteld door Aselect uit het gelijknamige rapport, Amsterdam, 1985.

Quist, J., *Duurzaam voeden in het huishouden, Toekomstbeelden & Analyseresultaten*, Document t.b.v. Workshop Implementatie. T.U. Delft, 2000.

Rescon, *Het perspectief project 1$^1/_2$ jaar later. De stand van zaken bij 11 huishoudens*, Rescon, Haarlem, 2000.

Schmidt, T.and A.D. Postma, *Minder energiegebruik door een andere leefstijl?* Project Perspectief Eindrapportage. VROM, Den Haag, 1999.

Schweitzer, R. von, *Überlebt der Haushalt das Jahr 2000?* Deutsche Gesellschaft für Hauswirtschaft, Essen, 1975.

Siderius, H-P, P Hedenskog, L. Sillanpää, Washing machines, driers and dishwashers, Basic assumptions, test methods and consumer aspects, Background report I, Group for efficient appliances, Working group – European Energy Network, Danish Energy Agency, 1995.

Vanek, J., Time spent in housework, *Scientific American*, Vol. 231, nr 5, 1978.

Uitdenbogerd, D.E., N.M. Brouwer, J.P.Groot-Marcus, *Domestic energy saving potentials for food and textiles. An empirical study*, H&C onderzoeksrapport 2, Household and Consumer Studies, Agricultural University, Wageningen, 1998.

Wal, J. van der and H.C. Moll, Towards sustainable household energy use in The Netherlands. *International Journal of Environment and Pollution*, Vol. 14, 2001, pp. 217-230.

Wilting, H.C., H.C. Moll, S. Nonhebel, *An integrative assessment of greenhouse gas reduction options*, IVEM onderzoeksrapport no. 101, Groningen, 1999.

Chapter 6

The role of local authorities in a transition towards a climate-neutral society

Frans Coenen and Marijke Menkveld

In this chapter...

... the role of local authorities in a transition towards a climate-neutral society will be elaborated. This is a level of government action that often tends to be forgotten when talking about global climate change. Local authorities operate in very different playing fields, which differ between countries and can change over time. The role of government can be relatively reserved or can involve a more active role by local authorities in regulation, spatial planning, sustainable housing and sustainable production and consumption.

Efforts to enlarge the municipal contribution can follow two paths. The first path is to increase the role of local authorities by expanding their competence and playing field. The second path is to use the existing municipal playing field as advantageously as possible. Many local climate options are known in theory. Unfortunately, as evaluation of the local climate options shows, many options for climate reduction are not used even when municipalities have the competence to use them. Expanding the role of local authorities would be asking them to play their part in problem solving.

Introduction

Although climate change is first and foremost a global environmental problem, it requires action at all levels of government. Primary responsibility for the implementation of international commitments, like the Kyoto agreements, lies with the national governments that signed them. The European Commission is a separate signatory. In Europe, these two levels of government are crucial for climate change policy because they set the appropriate climate change strategies.

However, it is clear that climate change has a local dimension as well. This chapter discusses the role of local authorities in a transition towards a climate-neutral society. Literature and policy practice throw up several general reasons why local authorities are important to climate policy. First of all, as the layer of government that is closest to the citizen, local authorities can play a major part in creating a degree of support for climate policy. Secondly, local authorities are responsible for local activities that result in greenhouse gas emissions, ranging from public lighting to the energy consumed in public buildings. And thirdly, they have responsibility in many areas of policy that are significant for climate policy, such as town and country planning, construction and transport. These responsibilities are not just very important for their own, local climate policy; they are also significant for the role of the local council as co-executive of national and international climate policy.

Given their position as the level of government closest to the citizen and their limited, but crucial, competence in important areas of climate policy, in this chapter we pose the question: *what can and should local authorities do to contribute to an 80 per cent greenhouse gas emission reduction?*

The answer to this question depends on the particular national, political and institutional context in which these local authorities have to operate. The structural features of this context play a significant role in how (local) climate policy is adopted and then put into effect. The substance of local climate policy depends on the specific nature of the community in question, such as its geography, demography, economy, society and culture (compare Lafferty and Eckerberg, 1998). We call the framework in which a local authority has to operate its *playing field*. For the answer to the question of what the role of local authorities can be in a transition towards a climate-neutral society, it is crucial to realize that this playing field differs between countries and can change over time.

In the first place, municipalities in different countries around the world have different playing fields as regards their powers to steer and their economic possibilities to implement measures. Given these differences, one cannot simply conclude what municipalities should do from international documents. Supranational initiatives leave considerable room for cross-national variation to reflect the diversity in institutional and political local context.

Second, the playing field for local authorities is not static. For instance, the two visions of how emissions could be reduced by 80 per cent, described in Chapter 2 of this book, contain different assumptions about the role of local government in the future (see below).

Because of the specific characteristics of countries, it is difficult to define a general playing field for local authorities and, perhaps even more importantly, to define what needs to be changed in this playing field to optimize the contribution of local authorities to a trend break towards a climate-neutral society. There are two options that could potentially enlarge the municipal contribution. The first option is to increase the role of local authorities by expanding their competence and playing field. The second is to use the municipal playing field as advantageously as possible.

Option one, an enlarged role for local authorities, assumes a change in the role of local authorities in climate policy. In the next section we will discuss the role of local authorities in climate policy, the factors that determine this role and the potential changes in this role. Option two, optimizing the use of the existing playing field, is specific to each country and local authority. We will discuss this trend break using The Netherlands as an example. Then we will discuss the role of local climate policy in The Netherlands, the local climate options and their total potential for greenhouse gas emission reduction and the use of their playing field by the Dutch municipalities.

Finally, in the last section of this chapter we will draw some conclusions about the potential role of local authorities in a transition towards a climate-neutral society.

The role of local authorities

Factors determining the role of local authorities in climate policy

We can address the role of local authorities in a change towards a climate-neutral society from both a normative and an empirical perspective. Empirically, we can ask what is the role of local authorities in climate policy and what factors determine this role. Normatively, we can ask what this role should be and according to what criteria.

Climate change is a classic example of multi-level decision-making (Collier, 1997). At the highest level of decision-making we find the United Nations, with the Framework Convention on Climate Change (UNFCCC) and the Kyoto Protocol. At the supranational level, in Europe the European Union influences climate change through its energy-efficiency, renewable energy and energy R&D activities, the EU structural funds, the common agricultural policy and the internal market. At the national level, national governments bear responsibility for economic, energy and transport policies.

Although there are many differences between countries, local authorities are generally responsible for local transport planning, spatial and development planning and energy management. Last but not least, there is the level of the citizens and communities where decisions are made about the ultimate use of energy and choices are made about transport and the use of renewable energy.

In general, the powers of local authorities with respect to energy and climate matters vary greatly from one country to another. By analogy with Local Agenda 21 implementation, one would expect that a high degree of autonomy for local authorities coupled with a history of environmental concern and actions will provide a solid basis for local climate policy. One would expect this to depend not only on the degree of local autonomy in general but also on (1) the specific scope of autonomy with respect to local climate policy, and (2) the local capacity for action in terms of the resources and competencies of local authorities (compare Lafferty and Eckerberg, 1998, pp247–249).

In recent years several studies (Energie-Cités/Ademe Franche-Comté, 1996; Energie-Cités/Fedarene/Islenet, 1996; Climate Alliance, 1995; Collier, 1997) have been published comparing the situation in the field of energy policy and climate policy in different European countries.

Energie-Cités, an association of European municipalities that promotes sustainable local energy policies, published a report (Energie-Cités/Ademe Franche-Comté, 1996) containing fact sheets on energy and municipalities in 15 European countries. The report identifies several important factors that explain the diversity between the countries investigated. First, the history and the resulting administrative organization of each country have determined the relative weight of local power. In each country the balance between national and local powers will determine the general room to manoeuvre enjoyed by municipalities, particularly in the energy sphere. Second, the presence, or otherwise the absence, of national energy resources has played a part in shaping energy systems and customs. And third, local climatic conditions have favoured intensive recourse to particular modes of energy: the production and distribution of heat is naturally more developed in countries where thermal needs are particularly acute. A fourth factor is the political circumstances, which have often had a strong impact on energy systems. Examples include the nationalization of gas and electricity at some point in the past in certain countries, the start of the liberalization of the energy market in other countries or the decision by countries to abandon nuclear energy.

The report concludes that a map of Europe illustrating the powers and jurisdiction of municipalities in energy matters would broadly distinguish two groups of countries:

- countries in which the municipalities have considerable or very considerable powers: Sweden, Finland, Denmark, The Netherlands, Germany, Austria and (parts of) Italy. Urban energy planning and a truly municipal energy-management policy are often to be found in these countries;
- countries in which municipalities have few or limited powers: the United Kingdom, Ireland, France, Belgium, Spain, Portugal and Greece. Situations may vary within this group, but very few municipalities possess an urban energy planning system.

A project carried out at the European University Institute concentrated more specifically on climate change policies, including an examination of local action in five EU member states (Collier, 1997). This report concludes that climate change activities can be expected to depend on two main factors. First, the competencies in crucial areas of energy and transport, which affect the scope of actions by local authorities. Second, public sensibility both to environmental issues in general and climate change issues in particular. Collier suggests that local authorities are unlikely to get involved unless there is public and political support. On that basis, she expects more action from so-called 'leader' countries.

Together with a number of partners, the Climate Alliance (an association of concerned local authorities, see below) conducted a survey of municipal activities to reduce CO_2 emissions and of the national framework conditions for local action in four European countries (Austria, Germany, Italy and The Netherlands). The national framework conditions were analysed to identify obstacles to local climate policy. For the purposes of this chapter, the overall conclusions concerning the role of the national framework are particularly interesting. The report concludes that the influence of the national framework is highly relevant to the extent and intensity of municipal climate protection activities. However, in some cases there are severe legal obstacles to further activities, while in most cases the competencies of municipalities are not clearly defined or there is a lack of funding or know-how. Even under the existing framework conditions there are many options for municipal carbon-reduction strategies and measures that no municipality has actually completely exhausted. Nevertheless, local authorities could act much more effectively if the national and European framework conditions were set properly (Climate Alliance, 1995).

Collier (1997) reaches a similar conclusion on the importance of policy frameworks at the national and EU level. As 'unfavourable', she identifies budget constraints imposed on local authorities, the pursuit of low energy prices combined with a resistance to imposing carbon taxes, the absence of energy-efficiency standards and insufficient support for public transport.

To summarize the most relevant factors for our discussion here, the role of local authorities is determined by:

- the general autonomy of local authorities;
- their specific autonomy in crucial policy areas of energy and transport;
- public and political support for environmental and climate issues.

Appropriateness of the local level of governance in climate policy

Why is local government an appropriate level of governance for climate policy? In the introduction we gave three arguments for the importance of local climate policy. Two of the arguments given, that local authorities initiate local activities that result in greenhouse gas emissions and that they have responsibility in many areas of policy that are significant for climate policy, refer particularly to the criterion of effectiveness.

The EU captures the effectiveness criterion for the most appropriate level of governance for climate policy in the following formulation: *'to establish the level best suited to the type of pollution and the geographical zone to be protected'* (First EU Environmental Action Programme). This criterion refers to the extent that climate problems have their roots in local activities and local government's initiatives. Local policy therefore can be effective policy.

The argument that local authorities are the level of government closest to the citizen is in itself a normative criterion for local democracy. The normative criterion 'as close to the citizen as possible' is linked with effectiveness through public involvement and its consequences for 'raising awareness' and 'building support' for climate policy.

For instance, according to Chapter 28 of Agenda 21, local authorities can deal very effectively with public involvement because *'as the level of governance closest to the people, they play a vital role in educating, mobilising and responding to the public to promote sustainable development'* (Chapter 28, Agenda 21).

The question of why local government is an appropriate level of governance for climate policy brings us to the concept of subsidiarity. On the basis of the work in the SUSCOM project, Lafferty (2000) states that the principle of subsidiarity can be understood in two ways. One has its roots in a historical-philosophical discourse, where subsidiarity refers to the idea that superordinate decision-making bodies should only exercise powers that are ostensibly necessary for securing a 'better' (more effective, more efficient, more morally correct) state of affairs for individuals and institutions at a lower level of organization. The Maastricht Treaty states that: *'in areas which do not fall within its exclusive competence ... the Community shall take action ... only if and in so far as the objectives of the proposed action cannot be sufficiently achieved by the member states and can therefore, by reason of the scale or effects of the proposed action, be better achieved by the Community.'* The burden of proof in exercising this understanding of the principle is that the transfer of authority upwards gains its legitimacy from a necessary or desirable additional contribution that is not attainable through the efforts of lower-level domains.

Lafferty refers to the second connotation of subsidiarity that he distinguishes as more 'modern' and politically motivated. In his opinion, in many ways this interpretation turns the first connotation on its head. Instead of arguing why it is necessary and/or desirable to provide subsidiary authority at a higher level of governance, it is used to argue for the moral necessity of keeping decision-making authority as close to local decision-makers as possible. What was originally a principle to justify a transfer of 'sovereignty' upwards, here becomes a principle to enhance democracy and protect sovereignty at a lower level: 'Decisions should be taken as close as possible to the citizens affected by the decisions'.

If we apply the principle of subsidiarity to the dilemma of 'decentralization versus the necessity of national direction of the transition' we have to be aware that the subsidiarity debate focuses on the division of powers between the supranational bodies of the European Union and the decision-making bodies of the member states. Secondly, the two interpretations of subsidiarity are interrelated. As Lafferty states, there is implicit reference to the more philosophical, legalistic connotation in the more political connotation. The decision that is to be taken closest to the people affected must be the best decision possible within the scope and functionality of that particular unit.

The potential changes in the role of local authorities in climate policies

Local policy-making is important for realizing a radical change in greenhouse gas emissions because local policies are important for the key aspects of transition processes that:

- deal with the use of space;
- deal with the built environment;
- rely on household behaviour and consumption.

How important local policies will be in transition processes depends on how the 80 per cent greenhouse gas reduction goals are to be reached. In the two visions of the Dutch energy systems in 2050 (see Chapter 2), the role of government, and particularly of local government, differs quite appreciably.

In vision A the role of government will be more reserved. Vision B assumes a stronger government with greater regulation. In the internationally-oriented 'global village' in vision A, the role of local authorities will be smaller than in vision B, which is more regionally oriented. The role of local spatial planning will be quite different in the two visions. The assumption in vision A is that people live in large houses in spacious suburbs, while in vision B people live in compact cities in smaller houses. In vision B municipal land-use planning will be more restrictive. In vision B renewable energy is strongly supported. The municipalities have to reserve space for the wind-power capacity that has to be installed on land. Furthermore, local authorities have to create the conditions for large-scale application of solar energy in the built-up environment.

As regards dwellings, local authorities bear responsibilities for energy efficiency, for instance through standards of insulation. In vision B the local authority can make use of spatial measures that will favour energy-infrastructure measures such as district heating, compact building, passive solar heating and the integration of combined heat and power (CHP) generation. In vision A a completely revised energy infrastructure will be needed.

In vision B, public transport also plays a much larger role. Municipalities can play an important role in stimulating people to get out of their cars and into public transport. Finally, local authorities can play an important role in raising public awareness and creating support for reduction of greenhouse gas emissions. In vision B, consumer behaviour is itself an essential component of reducing greenhouse gas emissions.

To conclude, vision B envisages a more active role by local authorities in regulation, spatial planning, sustainable housing and sustainable production and consumption.

Dutch municipalities as an example

The role of local authorities in Dutch climate policy

Because it is difficult to define a universal playing field, we will take a specific country as an example. Given the factors determining the role of local authorities in climate policy discussed in the previous section (general autonomy, specific autonomy and public and political support for the climate) we will briefly discuss the Dutch situation, applying the factors discussed above to the Dutch example.

General autonomy of the Dutch local authorities

Classifications of 'models' or types of local government systems (Hesse and Sharpe, 1991) place The Netherlands within the 'Napoleonic' system. This model is characterized by a history with a relatively high degree of centralized state control. Local authorities tended to be 'supervised' by mayors, who were in turn strongly associated with the central government. The Dutch constitutional system is generally labelled as a 'decentralized unitary state'. The unitary nature of this type of state is based on agreement between the three levels of government (national government, provinces and municipalities) and not on central government. Local authorities have freedom of initiative within the system of inter-administrative relations. They have the constitutional power to deal with matters of local concern as long as the local authorities take account of legislation passed by higher authorities. Many local authority activities are covered by legislation within a system of co-government.

Dutch local authorities have relatively limited tax revenues of their own and depend largely on central government for their resources. About half of this central funding takes the form of a specific transfer, or earmarked funding. The rest of the central funding is in the form of a Municipal Fund, which is an open-budget system with a budget ceiling, with its distribution depending on criteria such as the number of inhabitants.

Public and political support and specific autonomy

The Netherlands can be regarded as being among the 'leader' countries in environmental policy (Jänicke and Weidner, 1997; Organisation for Economic Co-operation and Development, 1995). Most Dutch municipalities have had an energy policy since the 1970s. In the early period the main aim of this energy policy was to conserve energy for financial reasons, given the rising energy prices. Energy saving was an aim in itself. Very little attention was given to the role that municipalities could play with respect to other energy users like businesses and citizens. From the end of the 1970s onwards there was a lot of discussion between national and local government about the way municipalities performed their local environmental tasks. The major problems were the lack of priority given to local environmental policy in general and the lack of local capacity in terms of resources and manpower. As a result of these discussions, during the 1990s municipalities received earmarked funding to bring their environmental policy to a basic and adequate level.

In the first National Environmental Policy Plan (NEPP) (1989) and the first white paper on energy saving (1990) municipalities were given a role as co-executor of energy policy. Energy policy was linked to sustainable development through key themes like the greenhouse effect, acid rain and the depletion of resources. To stimulate the implementation of the first NEPP, the ministry of the environment and the umbrella organization for local authorities (VNG) produced a framework for the implementation of the NEPP by municipalities (1990). This framework included basic mandatory and voluntary tasks in the field of energy, including the drafting of a municipal energy plan. NOVEM, an agency largely funded by the Dutch government which manages many national and international governmental programmes aimed at environmental improvement, increasing energy efficiency and introducing renewable energy sources, stimulated the drafting of these plans by subsidizing municipalities if they adopted a municipal energy-saving programme (known in Dutch as the GEA method). From 1 January 1995 the first earmarked funding programme, the so-called BUGM[1] funding, was followed by an extended programme, the VOGM[2] funding, which ran until the end of 1997. In this latter programme, energy saving could be chosen as a voluntary priority task (known in Dutch as Ebb, which stands for energy-saving policy) Almost two-thirds of all municipalities joined this VOGM programme, which incorporated the GEA municipal energy-saving approach.

During 1995, however, the national government and the Climate Alliance identified a serious bottleneck; many municipalities were experiencing great difficulties in the implementation of the formulated policy due to obstacles in terms of finances, political commitment and organizational match.

At that time more than half of the municipalities had drawn up an energy-saving plan using the GEA method. This led to a new initiative, the so-called LOREEN[3] programme, which was specifically designed to support municipalities and remove obstacles to the municipal implementation process. After the period of earmarked funding, many local governments had to improve their environmental policy and further extend their own responsibility.

When the new agreements were being concluded between national and local government, climate change was raised as an issue. All parties are now working on a so-called climate voluntary agreement. Many municipalities are interested in concluding an agreement on municipal climate policy with the national govern-

---

[1] BUGM in Dutch refers to a government-sponsored programme, `Funding implementation municipal environmental policy'.
[2] VOGM in Dutch refers to a Government-sponsored programme for 'Continuation of support of municipal environmental policy'.
[3] LOREEN in Dutch refers to 'Programme for Regional and Local Energy Saving'.

ment in which municipalities will agree to give climate policy a prominent place on the local policy agenda in exchange for national government support (financial resources, knowledge, adequate legislation, etc.).

At the time of writing, about two-thirds of all municipalities have adopted the GEA approach. A survey shows that 70 per cent of these municipalities, for whom the planning period is coming to an end, have plans to develop follow-up policies (NIPO, 2000). In addition, nearly all municipalities that formulated policies are actually implementing them or are already in the phase of formulating new policies (NIPO, 2000). The willingness of Dutch municipalities to get involved in climate policy is also demonstrated by other initiatives. At the time of writing, one in six local authorities (around 100 in all) have joined the International Climate Union, an alliance of native populations in the Amazon basin and western local councils committed to the ambitious target of reducing CO_2 emissions by 50 per cent from 1987 levels by 2010. A quarter of the local authorities in The Netherlands have responded to the call in Chapter 28 of Agenda 21 and set up a local Agenda 21.

The greenhouse gas reduction potential of Dutch municipalities

Municipalities have many options to reduce greenhouse gas emissions. The theoretical reduction potential of municipal policies depends on three factors; first of all, the extent to which a certain reduction option is within the area of competence of a municipality. This competence, as we have seen, will differ from country to country. Secondly, not all climate options within the competence of municipalities in a particular country are equally relevant to all local authorities. Some of the options are directly linked to certain characteristics of a local authority, such as having a new housing estate or an industrial estate or businesses with a certain level of energy consumption within its boundaries. Thirdly, there is the extent to which options are already used by municipalities.

In this section we will discuss the potential for greenhouse gas reduction by Dutch municipalities. A research project estimated the emissions of greenhouse gases related to activities that fall within the sphere of influence of Dutch local authorities to be more than 40 per cent of The Netherlands' total greenhouse gas emissions (Burger, Coenen and Menkveld, 2001).

The following summary outlines the most promising local options in five areas of local authority responsibility: town and country planning, construction and housing, traffic, environment, and local authority management tasks. The list of options given here is by no means exhaustive, but illustrates what can be attained and what is actually being achieved at the moment. In the following matrix the tasks are placed along one axis and and the target groups of climate policy along the other. The various climate policy target groups identified in the Dutch Environmental Balance (state of the environment report) under the heading of climate change are utility building (or: Trade, Services and Government), households, transport and traffic, waste, the energy sector, industry and agriculture. The local authority playing field encompasses a mixture of options and available instruments under direct municipal influence and indirectly via target groups on the one hand, and its areas of responsibility on the other. Table 6.1 presents a summary of

Table 6.1 Areas of local authority responsibility for various target groups

	Areas of local authority responsibility				
	Planning	Construction and housing	Transport and traffic	Environment	Local government tasks
Utility Buildings	compact building, passive solar energy/zoning plan energy infrastructure, energy vision	Sustainable Construction and Energy Performance Standard (EPN)/maintenance, information, covenants, etc.		energy saving and energy/environmental licence, environmental communication	energy saving, local sustainable government buildings/environmental care, subsidy conditions energy saving, public transport and traffic lights/public lighting plan
Households	compact building, passive solar energy/zoning plan energy infrastructure/ energy vision	Sustainable Construction and Energy Performance Standard (EPN) / maintenance, information, covenants, etc.		energy saving and sustainable energy/ environmental communication	
Traffic	decrease car mobility/ spatial planning, location policy/low traffic density neighbourhood/ zoning plan		decrease car mobility/ traffic planning, parking	transport management at companies/environmental policy cycle route plan	sustainable industrial estate improvement public licence transport transportation management own organization

Table 6.1 continued

	Areas of local authority responsibility				
	Planning	Construction and housing	Transport and traffic	Environment	Local government tasks
Waste	residual heat utilisation/ structure plan, choice energy infrastructure			waste prevention and recycling at companies/ environmental licence sustainable industrial estate land fill gas extraction and kitchen and garden waste fermentation/waste disposal	
Energy sector	residual heat utilization/ structure plan space for sustainable energy/ zoning plan	sustainable energy/ Sustainable Construction policy			
Industry	residual heat utilization/ structure plan, choice energy infrastructure			energy saving/ environmental licence sustainable industrial estate	
Agriculture	utilization CO_2 or residual heat/structure plan choice energy infrastructure space for sustainable energy/zoning plan			energy saving/environmental licence	

important local options for greenhouse gas reduction, given the possibilities that the local government has to influence them, in five areas of responsibility. We will briefly discuss these options, their reduction potential and how this potential has been exploited in practice up to now.

Town and country planning

Spatial planning can extend the possibilities for waste heat utilization by locating consumers of heat (for instance, a residential area) close to a heat producer (for instance, a power plant or a waste incinerator). This means that in the construction of the energy infrastructure in new or redeveloped residential locations, choices have to be made regarding the heat-distribution network. The Dutch national government expects a substantial contribution to CO_2 reduction from the use of residual warmth and the application of combined heat and power (CHP). The point of departure for the national government is that installed CHP capacity should double between 2000 and 2010 from 8000 MW to 15,000 MW, which could represent a total CO_2 reduction of 5-10 Mtonne CO_2 (Uitvoeringsnota Klimaatbeleid, 1999).

At the moment, the availability of waste heat is scarcely considered in decisions on the location of potential consumers of that heat, such as new residential areas, industrial areas or agricultural greenhouses. The subject is mentioned in national and provincial spatial planning documents, such as the Fifth National Spatial Planning Framework, but in practice not much happens yet.

Van der Waals et al. (2000a) investigated the decision-making processes surrounding the design of 26 larger new residential areas, the so-called VINEX areas, with more than 4,000 houses. In 15 of the 26 locations, a 'heat network' was either chosen or a strong preference was expressed for such a network. The possibilities in smaller new housing areas are expected to be less positive. The areas covered by the study represent 226,000 houses already built or to be built between 1995 and 2005. The total number of new houses built in The Netherlands each year is about 100,000, so the areas studied represent about a quarter of the total number of new or planned houses. Urban planning instruments like compact building and passive utilization of the sun (orientation of plots to the sun) can also contribute to CO_2 reduction. The same study shows that in many of the VINEX locations more than 50 per cent of the plots are oriented towards the south.

Municipalities play an important role in creating space for wind energy. Wind energy with a primary energy-saving potential of about 50 to 100 PJ (CO_2 reduction of about 3 to 6 Mtonne) could become one of the most important sustainable-energy options in the longer term. Unfortunately, few Dutch municipalities have an active policy towards wind energy. So few, in fact, that the national government has threatened to force municipalities to install wind turbines.

In theory, urban planning could help to reduce car mobility. The impact of these measures is uncertain. People seem to be less sensitive to distance. The potential reduction of car mobility in any urban area is limited since it doesn't depend on the number of car movements but on the number of kilometres travelled by car. Only about 6 per cent of all kilometres travelled in a car involve journeys over a distance of 5 kilometres or less, journeys that could be made by bicycle. However, a housing area should offer the organizational infrastructure to make car-sharing, car-pooling and teleworking possible. The research referred to above shows that in all the VINEX locations studied, alternatives to using the car have been stimulated, but only in a few cases alternatives to car use or car ownership.

Construction and housing

The reduction potential of sustainable construction is large; 40 to 50 per cent of emissions by households and utility buildings are caused by the use of energy for heating. So if sustainable construction measures can reduce the energy use in buildings by 10 to 15 per cent, we are talking about a serious reduction potential of several Mtonne CO_2 per year.

In 1998, 32 per cent of all newly built houses complied with standards that exceeded existing legal sustainable-building requirements laid down in a sustainable construction performance standard. The extra measures involved the application of all cost-neutral or cost-saving measures set out in the National Package Sustainable Building and all variable measures involving extra investment of up to 2,000 guilders (Beleidsprogramma 2000-2004).

In the existing housing stock, there were example projects for 56,000 houses in the period 1995-1998 and subsidies were promised for sustainable renovation. An inventory of energy saving in housing renovation plans covering about 70 projects, which involved around 1,000 houses, shows that about one-third of the plans included energy saving as a separate goal. Some smaller sustainable measures relating to buildings, like solar energy, solar boilers and heat pumps, are only used on a small scale or in experiments (Van der Waals et al., 2000b). All in all, we can conclude that measures for sustainable construction are frequently used, especially in one-third of new-built houses, but that it is certainly not common practice.

Transport and traffic

Options in traffic range from decreasing car mobility (encouraging cycling, car sharing, teleworking, transportation management) to encouraging public transport and car pooling to electric vehicles and biofuels. In the area of transport and traffic, traffic planning, spatial organization and parking policy are used to reduce car mobility. Other options are the insertion of transport management in environmental licences of businesses or in sustainable industrial estates.

Stimulating public transport is probably the most promising option in terms of reduction potential in this policy area. It is unclear how much influence municipalities have here. Municipalities can also stimulate car sharing, car pooling and Park & Ride options by creating the infrastructure.

Much of what is happening in this policy field is targeted at parking. However, the motivation behind parking policy is seldom climate policy but more to do with tackling congestion in inner cities. Municipal measures are concerned with improving the logistics of urban distribution.

Environmental policy

Municipalities can prescribe energy measures in the environmental permits of small businesses and the service industry. National government is responsible for arrangements with the energy-intensive industries through a voluntary agreement, the so-called Convenant Benchmarking, which is expected to cover about 80 per cent of all industrial energy use. However, industry in The Netherlands uses so much energy (1,100 PJ) that the remaining 20 per cent still represents a large part of the total energy use in The Netherlands and accounts for about 5-10 per cent of all greenhouse gas emissions. Furthermore, the service sector is also responsible for about 13 per cent of all greenhouse gas emissions in The Netherlands. According to the national government, energy saving has not received enough attention in environmental permits issued by municipalities in recent years. Energy-saving measures in permits go beyond just applying energy measures in the production processes. They also relate to indirect energy savings by reducing mobility, the economical use of resources and waste prevention.

Dealing with traffic management in permits is difficult in practice, although the reduction potential is large. One-third of all kilometres travelled by car involves commuter traffic. Traffic management could reduce this number by between 6 per cent and 20 per cent. Traffic management can also be applied to the transport of goods by firms.

Waste prevention has been a topic of discussion in The Netherlands for many years and a great deal of experience has been gained. An important driver for waste prevention in (larger) businesses has been the 'pollution prevention pays' principle. Environmental permits do not devote much attention to waste prevention and resource use although municipalities do have the legal power to include them. There is a reduction potential here for the stragglers and companies without environmental care systems.

Separate waste collection is a crucial condition for recycling. In 1995, 42 per cent of domestic waste was collected in separated streams and the goal is 60 per cent by 2000. Especially the recycling of products from energy-intensive industry (like metals, paper and glass) contributes to energy savings in production, while recycling of synthetic materials helps to curb the use of oil products and CO_2 emissions.

Methane emissions from landfills still contributed 4 per cent of the total volume of greenhouse gas emissions in The Netherlands in 1998, despite the use of landfill gas extraction. The use of landfill gas has a large reduction potential which could be used more, but is decreasing because of the separate collection of kitchen and garden waste which removes organic waste from the waste being dumped.

Local authorities also play an important role in communication about the environment. Climate change could be an aspect of environmental communication in raising awareness that climate policy and energy saving is necessary and that the target group has a crucial role to play in it. Local authorities can also inform climate policy target groups about laws, subsidies, etc. Finally, the municipality could communicate the example it sets in the field of energy saving.

Municipal management
A municipality can take measures in municipal buildings (swimming pools, sports facilities, theatres, offices, etc.). The reduction potential is limited. It is estimated that only 6 per cent of the energy use and emissions of public buildings can be attributed to municipal buildings.

There are many options for energy saving in public lighting (street lights, traffic lights). Public lighting uses 1,000 million kWh per year, and three-quarters of this is the responsibility of municipalities. It is believed that about 300 million kWh (1.5 Mtonne CO_2) could be saved at the municipal level. It is difficult to get an overall picture of how many energy-saving measures are already applied by municipalities. Many municipalities seem to have experience with one technical option and not others, but here also options don't seem to be commonly applied.

Lessons to be learned from the Dutch example

The use of the playing field
One could ask whether municipalities in a country are exhausting their climate options. A Dutch research project postulated the assumption that local government in The Netherlands could play a greater role than it currently does in reducing the country's greenhouse gas emissions.[4] Many local climate options are known in theory. Unfortunately, evaluation of the climate options identified in this chapter shows that many options for climate reduction are not availed of even though municipalities have the competence to employ them. Actual practice turns out to be rather fractious.

The evaluation research shows that many options for greenhouse gas reduction do not become common practice (Burgers et al., 2001b). An explanation lies in the obstacles faced by local authorities that want to implement these options:

- political barriers; a lack of political support for a particular policy option;
- communicative barriers; sometimes the policy direction or ambitions are unclear;
- organizational or institutional barriers, such as the lack of co-operation between different departments in the same local authority or dependence of local authorities on co-operation with other actors;
- resource barriers, such as manpower and finances, and the low expectations of municipalities for certain options.

Since 1996, the LOREEN programme has provided considerable support for Dutch municipalities that applied for it in overcoming these barriers to local energy and climate policy. Evaluation of this stimulation programme shows that development and implementation of both energy and climate policy improved considerably in these municipalities (Coenen et al., 2001). However, what is evident from the brief review of the climate options is that many of the options for local climate policy in fact fall outside the area of environmental tasks. This touches on the dilemma of 'transition through co-benefits' versus 'inevitability of a special climate change approach' and the dilemma of whether to separate climate change from sustainable development policies.

Furthermore, the possibilities for a specific target group to reduce greenhouse gas emissions always turn out to be divided over a number of areas of responsi-

---

[4] This chapter partly builds on a research project conducted for the Dutch National Research Programme for Global Air Pollution and Climate Change (NRP-MLK) that specifically addresses the potential role of the local level. The research project 'Local Government and Climate Policy' is a co-operative project between The Netherlands Energy Research Centre (ECN) and the Centre for Clean Technology and Environmental Policy (CSTM) at the University of Twente. The motivation for this research project is the assumption that local government in The Netherlands could play a greater role than they currently do in reducing greenhouse gas emissions in the country. This fact formed the motivation for initiating the research project on Local Government and Climate Policy.

bility, which leads to the conclusion that local climate policy is pre-eminently an integrated policy. If such an integrated policy is to be implemented, then institutional barriers within the local authority will have to be overcome and local climate policy will have to be approached more as an integrated problem extending also to non-environmental policy areas.

Potential for change

In the introduction we referred to two options for a trend break in the contribution of local authorities to climate policy. The first option is to enlarge the role of local authorities by expanding their competencies and playing field. The second option is to optimize the use of the municipal playing field. These options accord with the results of international research.

As we saw in the second section of this chapter, many of the cross-country comparative research projects look for factors in national policies that hinder local climate policies. Clearly these factors exist. In The Netherlands, the role of the local authorities in Dutch climate policy and climate research[5] has hitherto remained somewhat under-illuminated. For instance, little attention is paid to the local level in the national climate strategy as set out in the Climate Policy Implementation White Paper (1999).

Furthermore, the playing field is not static and the possibilities for local climate policy can change due to developments in society at large. The liberalization of the energy sector, the trend towards more decentralization and local standard-setting in environmental and energy policy and the changes in local democracy are particularly important for local climate policy in The Netherlands. The overall conclusion from research (Burger et al., 2000) that investigated how these developments may change the municipal playing field was that through the influence of societal changes the role of local authorities will become more important. Public expectations are higher and through processes of liberalization of the energy markets the co-operation with the profit-based energy firms will change. Processes of decentralization will mean that local authorities can rely less on national government. All these developments point to a greater role for local authorities in climate policy in the near future.

Trend breaks and dilemmas

In this chapter we considered two options for a trend break in the contribution of local authorities to climate policy: (1) enlarge the role of local authorities by expanding their competence and playing field, and (2) optimize the use of the municipal playing field.

We feel that such a trend break is essential for a transition towards a climate-neutral society. An overall reduction of emissions by 80 per cent can never be

---

[5] Until now only one of the Dutch National Climate research projects was explicitly concerned with the role of local authorities in climate policy. This project, done by the Amsterdam City Council with ECN, was especially concerned with setting out policy options for three activities: work, housing and transport. Little attention has been paid to the factors underlying the success or failure that the council may encounter when implementing these options. Several projects in the field of the built-up environment have of course a strong local dimension and study the role of local authorities in local planning processes (Van der Waals et al., (2000a,b)

achieved without a prominent role for the local level of governance. The Dutch example shows that 40 per cent of total Dutch greenhouse gas emissions are related to activities that fall within the Dutch local authorities' sphere of influence. This percentage is probably lower than in other countries, where the general autonomy of municipalities, and their specific autonomy in the fields of energy and climate policy, are larger.

The first trend break option is to enlarge the role of local authorities by expanding their competence and playing field. With respect to the specific dilemma of decentralization versus the need for national direction of the transition introduced in Chapter 3, it is inherently doubtful whether local government is an appropriate level of government for climate policy.

In the second section of this chapter we said that arguments in favour of a greater role for local authorities in climate policy could be based on both legitimacy and effectiveness criteria:

- the legitimacy criterion is that climate change policies should be developed and implemented as close to the citizens as possible;
- the effectiveness argument is that a large part, 40 per cent in the case of The Netherlands, of total greenhouse gas emissions are related to activities that fall within the sphere of influence of Dutch local authorities. This illustrates the fact that many climate problems have their roots in local activities and local governments' initiatives;
- this influence could be increasing given developments in society at large. In particular liberalization of the energy sector, the trend towards further decentralization and the changes in local democracy will cause the role of local authorities to change in the coming years. The role of local authorities will become more important and local authorities will therefore have to be more active.

As shown, a lot of attention is given in international literature to the national factors that may hinder local climate policy. The suggestion is that by removing these obstacles local authorities could act much more effectively if the national and European framework conditions are set properly. However, there are some serious dilemmas in striving towards a larger role for local authorities. Any form of decentralization of climate policy or giving local authorities a larger responsibility poses problems. Climate change is a difficult political issue at the local level because of two dilemmas mentioned in Chapter 3. The long-term perspective needed in climate-change policies is difficult for politics, which focuses more on urgent short-term local priorities. The relatively short-term character of local politics and the focus on urgent local priorities make it difficult to separate climate change from sustainable development policies. Furthermore, these characteristics of local politics not only make it difficult to achieve radical change, but also make it also difficult to prepare for radical changes in 30-40 years.

The second option, optimizing the use of the existing playing field, is specific to country and local authority. We discussed this trend break on the basis of the Dutch example. We sketched the role of local climate policy in The Netherlands, the local climate options of Dutch municipalities, their total potential for reducing greenhouse gas emissions and their use of the playing field.

This option would involve optimizing the contribution of local authorities within the existing national framework, presuming a large willingness to act among the majority of municipalities. A lack of competence should not be confused with an unwillingness to act. We believe that a greater contribution by local authorities should not be seen solely from the perspective of hindrances from outside the organization or of a lack of willingness to act, although both problems exist, but that of an optimal use of local climate policy options.

To enhance the contribution of local governments to climate policy, a systematic integrative approach is necessary. Evaluation of the experiences with local climate options and of the growing significance of local climate policy suggests the need for a larger contribution by local authorities. Improvements should be expected not only from climate options themselves but also from an organizational and process perspective. The local level is the pre-eminent level where policy decisions find concrete implementation and thus have a direct impact on daily practice. When policy decisions are being made it must rapidly be made clear what the consequences are for the emissions of greenhouse gases. This must make an integrated climate policy possible, piggy-backing with many other policy areas. Such an approach needs several basic steps or elements, which build on basic steps from total quality management and environmental care systems:

1. a policy document that articulates the political commitment to local climate policy;
2. plans and programmes to implement climate policy inside and outside the organization;
3. integration of these plans in daily policy practice and in the organizational culture;
4. monitoring the performance of local climate policy;
5. offering education and training to raise the awareness and understanding of the climate problem;
6. publishing information on local climate policy performance.

Following these basic steps would do justice to the potential options in climate policy and integration in other policy areas. Research (Menkveld et al., 2001) shows that the third step is the most crucial one in introducing 'climate care' throughout all tiers of the local authority. However, political commitment and drawing up plans are preconditions for integrating these plans in daily policy practice.

To answer the question of what local authorities can and should do to contribute to a 80 per cent greenhouse gas reduction, we suggested two pathways. The first path is enlarging the role of local authorities by expanding their competence and playing field. The second path is using the existing municipal playing field as advantageously as possible. Expanding the role of local authorities would come down to asking many local authorities to do their bit for problem-solving. The observation that many municipalities only pursue some of all the possible climate policy options addresses the dilemma of a 'small numbers of decision makers' trajectory versus a 'decentralized multiple and diffuse decisions' trajectory. One strength of a larger role for local authorities could be wider public support because of the smaller distance between local politics and local people.

Local climate policies encompass a broad range of policy fields. Expanding the role of local authorities in climate policy touches on the dilemma of 'transition through co-benefits' versus 'inevitability of a special climate change approach'. The relatively short-term character of local politics and the focus on urgent local priorities make it difficult to separate climate change from sustainable development policies. These characteristics of local politics also make it difficult not only to achieve radical change in the short term but also to prepare for radical change in 30-40 years.

References

Agyeman, J. and Evans, B. (eds.), *Local environmental policies and strategies*, Longman, Harlow, 1994.

Burger, H. *Evaluatie klimaatverbond*, Petten, 1999.

Burger, H. F. Coenen and M. Menkveld, *Het speelveld van lokaal klimaatbeleid*, rapport Fase 1 van het NOP-project Lokale overheden en klimaatbeleid, Petten, 2001a.

Burger, H. F. Coenen, M. Kaal and M. Menkveld, *Evaluatie van het speelveld van lokaal klimaatbeleid*, rapport Fase 2 van het NOP-project Lokale overheden en klimaatbeleid, Petten, 2001b.

Coenen, F.H.J.M., N.E. Marquart and H.G.J. Meynen, *Evaluatie programma Lokale en regionale energiebesparing*, Enschede, 2001.

Collier, U., Local authorities and climate protection in the European Union: putting subsidiarity into practice, *Local environment*, Vol 2. No 1, 1997, pp. 39-56

Energie Cités and Ademe Franche-Comté with the support of the European Commission, *The Powers and Responsibilities of Municipalities in the Energy Field*, December 1996.

Energie-Cités, Fedarene and Islenet, *Local and regional energy policies in Austria, Finland and Sweden*, 1996.

Hesse, J.J. and Sharpe, L.J, Local Government in International Perspective: Some Comparative Observations. In Hesse, J.J. (ed), *Local Government and Urban Affairs in International Perspective*. Baden-Baden: Nomos Verlagsgesellschaft, 1991.

Jänicke, Martin and Helmut Weidner (eds.), *National Environmental Policies: A Comparative Study of Capacity-Building*, Berlin, Heidelberg and New York: Springer, 1997.

Lafferty W.M., Subsidiarity in practice: problems and prospects, in: L. Hakkinen, *Regions-cornerstones for sustainable development*, Academy of Finland, Helsinki, 2000.

Lafferty W.M. and K. Eckerberg (ed), *From Earth Summit to Local Agenda 21, Working towards Sustainable Development*, Earthscan, London, 1998.

Menkveld, M., H. Burger, H. Heinink, M. Kaal, F.H.J.M. Coenen en K.A. van der Veer, *Lokale overheden en klimaatbeleid*, Eindrapport van het NOP-project Lokale overheden en klimaatbeleid, Petten, 2001.

Ministry of Housing, Spatial Planning and Environment, *First National Environmental Policy Plan* (in Dutch), The Hague, 1989.

Ministry of Economic Affairs, *Nota Energiebesparing*, Tweede Kamer, vergaderjaar 1989-1990, 21570, nrs. 1-2. SDU uitgeverij, The Hague, 1990.

OECD, *Environmental performance reviews: The Netherlands*. Paris, OECD, 1995.

United Nations, *Agenda 21*, 1993.

Van der Waals, J.F.M., et al., *Energiebesparing en stedelijke herstructurering*, VROM, Utrecht, 2000a.

Van der Waals, J.F.M. et al., *CO_2-reduction in building locations, A survey and three case studies about the role of options for CO_2-reduction in planning processes*, NRP Bilthoven, 2000b.

Wallstrom, M., *Introductory statement on climate change*, Commissioner for Environment, EP plenary session, Strasbourg, 6.10.99.

Zuidema, M. and Rosebeek, A., *Gemeenten onderzoek, de 3-meting*, NIPO, 2000.

Chapter 7

Improved material management as trend-breaking technology for reduction of greenhouse gas emissions

Marko Hekkert, Peter Groenewegen, Tom Kram, Robbert van Duin and Paulien de Jong

In this chapter...

... the contribution of material management to reducing the final demand for energy will be elaborated. Improvement of material management is one of the ways of reducing the use of fossil fuels, alongside a wide range of other options like improvement of energy efficiency, application of renewable energy and shifts in the use of fuel types (from high to low carbon content). Unlike the other options, improving material management is not commonly incorporated in national and international greenhouse gas reduction policies. This is remarkable since several studies indicate that this option has great potential and is often economically attractive. Hence the fact that improved material management plays an important role in visions of drastic emission reductions is understandable, but it can be regarded as a breakthrough in its own right. Furthermore, this chapter will discuss the shift from industrial production towards services and present a qualitative evaluation of sustainable product-service systems as an important direction in the transition towards a climate-neutral society.

Introduction

Chapter 2 described two visions for transforming the energy system of The Netherlands into a climate-neutral system by 2050. The two visions differ strongly in terms of the final demand for energy and the technologies needed to meet this energy demand. However, in both visions, measures to reduce the final demand for energy play a vital role in keeping the growth of demand for energy under control. Specific attention is paid to the role of material management in reducing this final demand for energy. This is also the main focus of this chapter.

Improving material management is just one way of reducing the use of fossil fuels, along with a wide range of other options, such as improving energy efficiency, greater use of renewable energy and shifts in the use of fuel types (from high to low carbon content). Unlike the other options, improving material management is not commonly incorporated in national and international policies for reducing greenhouse gas emissions. This is remarkable since several studies indicate that this option has great potential and is often economically attractive. Hence, the fact that improved material management plays an important role in the two visions outlined in Chapter 2 is understandable, but it can also be regarded as a breakthrough in its own right.

The reason improving material efficiency may lead to reduced greenhouse gas (GHG) emissions is given below. Figure 7.1 gives a simplified overview of the life cycle of materials in an economy. Raw material production, material production and product manufacturing require large amounts of energy. Together, these processes form the industrial sector. In 1995, the industrial sector[1] accounted for about 40 per cent of the global total primary energy use (Price et al., 1998). A limited number of materials account for the bulk of this energy use and the associated greenhouse gas emissions. Table 7.1 provides an overview of these materials and the emissions associated with production and waste handling for the year 1995 in the Western European situation.

Table 7.1 shows that the production of these bulk materials accounts for roughly 1 Gton CO_2 equivalents of greenhouse gas emissions, which is approximately one-quarter of the total greenhouse gas emissions for Western European.[2] If materials were used more efficiently, either in the product manufacturing stage or in the consumption stage of products, less material would need to be produced and therefore less energy would be needed in the raw material and material production stage.[3] Consequently, more efficient material management is likely to lead to reduced emissions of greenhouse gases.

Life cycle of materials

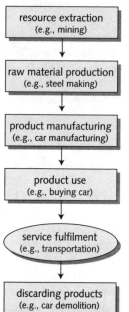

More efficient material management generally leads to *dematerialization*. There are different definitions of dematerialization in the literature but it generally refers to an absolute or relative reduction in the quantity of materials used in the production of a unit of economic output.[4]

In the next section, we will present some theoretical concepts to describe and analyse the relationship between material use and economic development, since this shows how necessary (and difficult) it is to strive for dematerialization. We then describe ways to use materials more efficiently and present modelling results on reducing greenhouse gas emissions.

Next we will consider a new trend (the shift from products to services) and discuss how this trend may affect material use and material policies. In the last sections we will focus on policy implications and end with some conclusions.

Figure 7.1 *Schematic representation of material life cycle*

---

[1] Excluding refineries.
[2] This excludes transport to consumers and product manufacturing from materials.
[3] It is also possible that other materials will need to be produced when materials are substituted. This leads to reduced greenhouse gas emissions when fewer greenhouse gases are emitted in producing the substitutes.
[4] Improved material management goes further than dematerialization. Material substitution, for example, is a valid measure to improve material management when it leads to reduced environmental impact. However, when it leads to an increase in material use (in kg), this measure is not regarded as dematerialization.

Table 7.1 *The annual emissions of greenhouse gases in Western Europe in 1995 due to the pro-duction and waste handling of materials, calculated according to accounting guidelines issued by IPCC (Hekkert, 2000)*

	CO_2 (Mt CO_2 equiv. pa)	Non-CO_2 greenhouse-gas (Mt CO_2 equiv. pa)	Total (Mt CO_2 equiv. pa)	Fraction (%)
Metals	244	11	255	26
Synthetic organic materials	167	53	220	22
Natural organic materials	93	130	223	22
Inorganic materials	49	60	109	11
Ceramic materials	191	-	191	19
Total	*744*	*254*	*998*	*100*

Some theoretical concepts to describe and analyse the relationship between material use and economic development

The materials that we use today are often more advanced than the materials we used in the past. Today we use high-strength steel to make lighter products, very thin aluminium drinking cans, high-strength concrete to build tall buildings, very lightweight carbon materials for numerous products and very small silicon chips with almost endless possibilities. It is no wonder, therefore, that people might think that advances in material technology take place autonomously and will eventually lead to very efficient systems of material production and use. This might even be regarded as a reason for not developing specific policy actions to improve the efficiency with which we use materials.

The literature offers some theoretical concepts that help us to understand the relationship between material use and variables like time, economic development or technological development.

In 1978, a study was published that posed a positive view of the sustainability of material use. In this study, Malenbaum (1978) examined the demand for metals in the US in the period 1951–1975. More specifically, he studied the *intensity of use* of metals, which can be expressed as the demand for materials in kilograms per dollar of gross domestic product (GDP). Malenbaum showed that for many metals, the intensity of use declined in the 1960s and 1970s. This led him to conclude that this development '...constitutes strong support for the argument that man's knowledge, skill and aspirations have served to slacken his need for industrial materials' (Malenbaum, 1978: 2).

Williams et al. (1987) continued the work of Malenbaum, studying the use of several basic materials over long periods, and concluded that the intensity of use of many materials declined over the periods studied. These developments can be depicted by means of a bell-shaped curve or inverted U-curve (see Figure 7.2). The bell-shaped curve can be seen as a decoupling of material use and economic growth. In this context, decoupling means that the intensity of material use declines while GDP keeps growing (dematerialization). The belief that the material intensity of use follows this kind of pattern led Williams et al. to state that

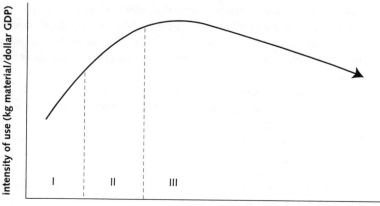

Figure 7.2 *Bell-shaped 'intensity-of-use curve', including a breakdown into three phases of economic development*

'...these trends mark the passing of the era of material-intensive production and the beginning of a new era in which economic growth is dominated by high technology products having low material content' (Williams et al., 1987: 140). These results are supported by Jänicke et al. (1989), who demonstrated a decoupling of material consumption (for selected materials) and economic growth for many nations. Roberts (1988, 1990) presented the same kinds of result for global and US metal use, respectively. In this period, many other studies, mostly on metals, showed similar results. See Cleveland and Ruth (1999) for an excellent overview of these studies.

The theory of why material use shows a bell-shaped pattern over time is as follows. When economies start to develop, material use is low and there are vast potential markets (phase I, see Figure 7.2). In this phase, material use grows rapidly in the course of building up an infrastructure (more rapidly than GDP). Economic growth enables improved education and understanding, which enables advances in processing technology, leading to improved quality of materials and further increases in demand (phase II). In phase III, the ratio of value-added to kilogram of material increases as more sophisticated materials are produced and innovations lead to more efficient use of materials. At this point, the demand for materials in kilograms per unit of GDP peaks and begins to decline. Later in this phase, markets for bulk materials become saturated and consumer preferences at high-income levels shift to less material-intensive goods and services (Williams et al., 1987).

In other words, these studies came up with results that confirm the idea put forward at the beginning of this section: that over time, materials will be produced and used in more efficient ways. This automatically leads to reduced environmental impact.

These optimistic results about the dematerialisation of economies were not shared by the entire scientific community. Cochran (1988) shows that the 'era of materials' has not passed at all because the volume (rather than the weight) of materials has increased dramatically over time. He states that expressing total material use in terms of volume rather than weight shows the switch from high-density materials like steel to low-density materials like plastics and aluminium in correct proportions. Other studies state that the measured trends in material use are not statistically proven and that no statements can be made about the future use of materials.

More recent studies are also not overly optimistic about the trends in material use. De Bruyn and Opschoor (1997) warn about overlooking substitutions between materials (just as Cochran did). They call this *transmaterialization* rather than dematerialization. Matos and Wagner (1998) also observed transmaterialization in material use in the US in the 20th century: the share of renewable materials declined from 45 per cent in 1900 to 8 per cent in 1995. De Bruyn and Opschoor (1997) found that dematerialization may have taken place in the last 20 years but that current trends show a recoupling[5] of material use and economic growth.

It is important to note that it is completely plausible for the intensity of use for a particular material or aggregate to decline while total use of materials increases because of rising affluence. In other words, less material is used relative to GDP but more material is used in absolute terms. De Bruyn and Opschoor (1997) therefore make a distinction between weak dematerialisation (a decline in intensity of use) and strong dematerialization (a decline in total material use). Wernick et al. (1996) note that population growth and rising affluence lead to increased material use, offsetting substitution, technical change and other forces that promote dematerialisation. Jänicke et al. (1997) show that for several countries, the demand for various materials did not decline even though the intensity of use generally did. Matos and Wagner (1998) also show that total material intensity declined dramatically in the US over the 20th century, while absolute total material use strongly increased.

So even though modern economies use materials more efficiently, the continuous growth of these economies leads to an absolute increase in material use, and it is therefore also likely that environmental impact increases.

To conclude, this overview of research on material use suggests a consensus that there is a declining intensity of material use based on weight in developing economies but that the absolute use of materials still increases. In other words, *strong dematerialization is a myth*. It does not take place. Breakthroughs are therefore necessary to reverse the current trend of weak dematerialization into strong dematerialization.

How can material-related greenhouse gas emissions be reduced?

The environmental pressure that is related to a product or a material is the sum of the environmental pressures related to the production of that material or product as well as the other stages in the life cycle of the product, such as production of raw material and waste management.

---

[5] Here, recoupling means that the material intensity is growing with GDP after a period where it has been declining in relation to GDP (decoupling).

Climate policy should therefore focus on greenhouse gas emissions at all stages of the product life cycle:

- raw material extraction
- material production
- product manufacturing
- product use
- disposal of products and waste management.

Changes in any of these stages can lead to reduced product-related greenhouse gas emissions. However, it is important to realize that measures that lead to reduced emissions in one part of the product life cycle may lead to increased emissions elsewhere: in order to reduce product-related greenhouse gas emissions, the whole life cycle needs to be taken into account.

Products are initially produced and consumed to fulfil a wide variety of functions; fulfilment of a function is the service that products deliver. For example, a car is bought to deliver a transportation service. Figure 7.3 gives an overview of the different life-cycle stages and presents an overview of measures that can be taken to improve the efficiency of material production and use.

Figure 7.3 shows that in all stages of the life cycle, measures can be taken to increase material efficiency. The improvements can be categorized in many different ways (see Worrell et al., 1995a and Duin et al., 1999). We use the following classification:

1. technical measures in the individual life-cycle stages
2. management of entire material product chain
3. sustainable product-service systems.

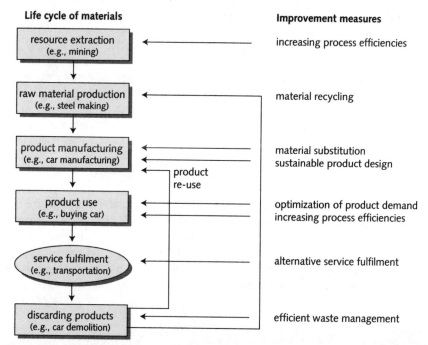

Figure 7.3 *Material life cycle and improvement measures*

Box 7.1.

Material substitution: the case of biomass

The use of biomass for material production may be a successful way to reduce material-related greenhouse gas emissions. The reason for this is that the greenhouse gas emissions associated with biomass are generally much lower than those associated with non-renewable materials. This is due to the CO_2-neutral nature of biomass (biomass takes up as much CO_2 during growth as it releases at the end of its lifetime). Two different types of biomass applications can be distinguished: the use of the structural capacities of biomass and the use of the chemical components in biomass.

The use of the structural capacities of biomass involves the use of round wood, sawn wood, and wood-based panels to replace materials like steel, concrete, and plastics. Several studies that focused on the construction sector have shown that this type of material substitution generates large reductions of greenhouse gas emissions. In Goverse et al. (2001), the authors calculated

that increased use of wood in construction of new residential buildings could reduce material-related greenhouse gas emissions by 50 per cent in this sector. This type of substitution is not only possible in the construction of residential buildings but also in renovations, construction of bridges, road applications (e.g., guard rails and road-sign portals) and non-residential buildings.

Biomass can also be used as feedstock in the chemical industry, thereby using the chemical components of biomass for the production of basic chemicals (so called C1-chemistry). In turn, these basic chemicals can be used for the production of plastics and other synthetic materials. Different routes can be distinguished for the conversion of biomass into basic chemicals. Biomass gasification is a method to produce syn-gas (a mixture of H_2, CH_4, CO, CO_2, H_2O), which can be converted into methanol (CH_3OH) used as feedstock in chemical production processes and as fuel.

Technical measures in the individual life-cycle stages include measures such as increasing the efficiency of processes (where innovations in technological development and process design can be used to reduce the amount of materials used in the process), improved product design (where using innovative production methods facilitates the use of thinner material [layers] and using other materials in the product manufacturing stage may lead to reduced emissions)[6] and more efficient waste management (where there is optimal energy recovery from waste materials).

Management of the entire material product chain covers measures like material recycling, product re-use (such as refillable plastic or glass bottles), material cascading (using discarded materials for the highest possible functions), extension of the useful life of products (because of modular design and increased fitness for repair and maintenance) and also improved waste management (e.g., choosing between incineration and material recycling).

---

[6] Important substitution criteria are the use of energy-extensive materials, materials with a long lifetime and materials that are suitable for re-use.

Sustainable product-service systems cover measures that lead to more efficient service because of the substitution of products by services or combinations of products and services. The idea behind this concept is that the same services can be provided with significantly less material.

The potential for measures to fall into the first two categories has been investigated in several projects. The results will be discussed in the next chapter. Far less research has been conducted into sustainable product-service systems, which will be discussed later in this chapter.

How much can greenhouse gases be reduced by improved material management?

Several studies have been carried out to determine the potential of material efficiency, (see Worrell et al. 1995a, 1995b; Fraanje, 1997; EPA, 1998). Most studies have focused on a single product group or material. Studies on packaging material by Hekkert et al. (2000a, 2000b) have shown that the greenhouse gas emissions associated with packaging material can be reduced by 40–50 per cent by means of material-efficiency measures. Studies by Worrell et al. (1995a, 1995b) have shown that more efficient use of fertilizers and plastic packaging could lead to significant reductions in CO_2 emissions (up to 40 per cent). Patel (1999: 217) has shown that a limited number of measures in the German plastics industry could lead to a 24 per cent reduction in carbon emissions. In his thesis, Gielen (1999) concluded that '...the potential for emission reduction in the materials system seems to be of a similar magnitude as the emission-reduction potential in the energy system'. In short, several studies have shown that the potential for reducing CO_2 emissions through improved material management is significant to say the least.

The problem with most studies that focus on a single material or product group is that they often overlook interactions between the material system and the energy system and interactions within the material system. In other words, energy-efficiency measures in industry have a negative effect on the greenhouse gas emission reduction of material-efficiency measures since less energy is associated with these materials. Moreover, thinner materials have a negative effect on potential substitution measures. A simple example: aluminium recycling to reduce primary production would not lead to lower greenhouse gas emissions associated with electricity production if all electricity were produced from renewable and/or nuclear resources. By contrast, the environmental benefit would be enormous if the electricity were produced from coal. In The Netherlands, a project has been conducted to model the Western European material system together with the Western European energy system in order to take the flaws of the earlier studies into account.[7] To underpin the model analysis, in-depth studies of the most relevant material and product groups were carried out (see, for example, Hekkert and Worrell, 1999; Hekkert et al., 1999; Bouwman and Moll, 1997; Joosten, 1998; and Daniels and Moll, 1997).

---

[7] This Dutch project was called the MATTER Project and involved The Netherlands Energy Foundation (ECN), Groningen University (IVEM-RUG), Free University (CAV-VU), Utrecht University (STS-UU) and Bureau B&G.

Box 7.2.
Improving the material life cycle: the case of re-usable packaging

Re-usable packaging is a typical example of improved material management with high potential for reducing greenhouse gas emissions. Re-usable packaging is common in The Netherlands and includes packaging such as bottles, boxes and carrier bags. Re-usable packaging implies that the packaging is used more than once, which saves on material use and waste production. This in turn leads to reduced emissions. Hekkert et al. (2000a, 2000b) calculate that re-usable packaging could lead to very large emission reductions compared to traditional packaging. Examples given by Hekkert et al. (2000a) and (2000b) are returnable PET bottles instead of disposable PET and glass bottles (-70 per cent and -85 per cent greenhouse gas emissions, respectively), Polycarbonate milk bottles instead of milk cartons (-15 per cent greenhouse-gas), returnable plastic crates instead of cardboard boxes (-35 per cent greenhouse-gas), and re-usable carrier bags instead of traditional plastic or paper bags (-96 per cent greenhouse-gas).

One needs to realize that from an environmental point of view, even though returnable packaging is a superb concept compared to disposable packaging, the implementation of this type of packaging in classic packaging systems is not easy. Hekkert (2000) describes this type of change as being very difficult since many actors are involved and the benefits are often not equally divided. In The Netherlands, this has resulted in several recent attempts by a number of industrial parties to abandon these re-usable packaging concepts in the Dutch packaging system. Fortunately, these attempts have been unsuccessful so far.

Figure 7.4 shows the contribution of individual material strategies to greenhouse gas emission reductions in Western Europe according to the model. It is apparent that the end-of-pipe strategy is important. The contribution of biomass as feedstock for the petrochemical industry is also significant (both fermentation and flash pyrolysis processes). The 'other resources substitution' in Figure 7.4 refers to the use of slag materials and Pozzolan for the production of cement and the use of tropical hardwood substitutes. There is some material substitution, such as the substitution of steel for aluminium in the transport sector and the substitution of concrete for wood in the building sector. Material substitution proves to be important because it induces emission reductions in the energy system (e.g., lightweight vehicles reduce fuel demand during transportation).

On the waste-management side, waste from renewable materials (wood, paper) is increasingly used for energy recovery, while waste from synthetic organic materials (plastics) is increasingly recycled. Waste management becomes especially important in the 200 Euro/t penalty case. Sensitivity analysis shows that the model would opt for disposal of these plastics as a carbon storage strategy if no restrictions were added. Improved quality (the category 'materials efficiency' in Figure 7.4) is introduced in the 200 Euro/t penalty case. One has to remember that the model representation of material efficiency is incomplete: the option has only been considered for steel and concrete because of the scarcity of reliable data for other materials. In addition, due to the character of the MARKAL model, cost-effective measures are part of the base case, and since most material-efficiency

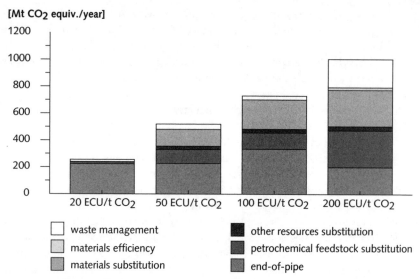

[Mt CO2 equiv./year]

Legend:
- waste management
- materials efficiency
- materials substitution
- other resources substitution
- petrochemical feedstock substitution
- end-of-pipe

Figure 7.4 *Contribution of individual materials strategies to emission reduction, 2030 (Gielen et al., 1999c]*

measures are cost-effective, they are not visible in Figure 7.4. As a consequence, the potential of material-efficiency measures in Figure 7.4 is severely underestimated. To illustrate how important this category is, in Hekkert et al. (2001) it is shown that the category 'material-efficiency improvement' contributes a 32 per cent reduction in greenhouse gas emissions compared to business-as-usual developments for the product group 'packaging materials'.

The model also shows that significant differences in consumption can be observed between the various groups of materials. Wood is a 'winner'. The significant increase in wood consumption is related to greater use of wood for buildings (coupled with less use of cement), for feedstock substitution and for paper. The increased wood use for paper production is related to increased energy recovery from waste paper. Another 'winner' in this analysis is aluminium, primarily due to increased use of it instead of steel in lightweight vehicles. Ethylene consumption is not affected strongly, but its production routes and waste handling are quite different (see Box 7.1).

The model results suggest that there is significant potential for reducing emissions on the side of material demand: at least 500–800 Mt CO_2 equivalents in Western Europe, which represents 20–35 per cent of total greenhouse gas emissions.

Sustainable product-service systems

There is evidence of a current shift in the basis of economic output from industrial production to services. This shift is generally considered to be beneficial for reducing environmental impact, including the climate problem, caused by economic growth. The shift to services is also felt to have significant consequences for the use of materials, but whether a shift to services will actually have a strong or weak dematerialization effect is a subject of debate. Models like the one described above do not provide answers to this question because of the lack of data on this subject. In this section, we will try to sketch a qualitative picture of the potential influence of sustainable product-service systems on material use and CO_2 emissions.

What are sustainable product-service systems?

The concept of sustainable product-service systems is a fairly new one. It is different from classic ideas like cleaner production and sustainable product design. Sustainable product-service systems are alternative socio-technical systems that may provide essential end-use functions, such as mobility and communication, that are normally provided by an existing product (Roy, 2000).

The key to sustainable product-service systems is that the eco-efficiency of customer activities is improved. This can be achieved directly by substituting an alternative product-service combination, but also indirectly by influencing customers to become more eco-efficient in their actions. In other words, there are material-related services and non-material services. The first category contains services that are developed in addition to a product: product-extension services (like repair and after-sales services), product-result services (where levels of performance are guaranteed by the supplier) and product-utility services (hire or lease contracts, for example). The other category of services is less product-oriented, for example, activity management services (optimization of processes), advice and consultation (energy or water efficiency), information provision (global positioning systems to control tractor spraying of fertilizers, for example), and intermediation (optimizing a match between supply and demand) and substitution possibilities (printing a tailored newspaper on demand, for example). These kinds of possible product-service system aim at goals of reducing CO_2 emissions similar to those described previously. Two shifts from products to services can be discerned:

- services as a substitute for, or in addition to, consumer products (this is often called product-services systems)
- services in (industrial) production chains. This trend is often referred to as the rise of the service economy

Services aimed at satisfying consumer needs

The discussion in environmental science on the connection between services and dematerialization has been stimulated by the work and findings of the Wuppertal Institute and Amory Lovins (Schmidt-Bleek, 1994). The material intensity per service unit (MIPS) is a central theme that has been developed by the Wuppertal Institute. This approach claims that services are the central element for fulfilling human needs. In order to provide needs such as food, heating and clothing, various options are possible in different combinations of products and services. An optimal service from an environmental perspective is one that uses the least amount of material and energy to fulfil a need. In their Factor Four book, Schmidt-Bleek (1994) show many examples of how services may lead to a material-extensive method of consumption. A well-known example that has been successfully applied in practice is car renting. The fact that several persons share the use of a car results in significant reductions in material input for car manufacturing. It is suggested that the kind of products that are best suited for this kind of service include those with high durability (which allows them to be used by a series of customers), low utilization rates, and rapid obsolescence.

Services in industrial production

The efficiency of organization in the production of consumption goods determines the energy and material intensity of production. With the application of information and communication technologies (ICT), new relationships in production structures are possible, which may lead to more efficient production. One of the ongoing developments in the change in relationships between different parts of the central production organization is outsourcing, where elements of production are handled outside the core firm.

For instance, if a manufacturing company handled the cleaning of production equipment and company premises itself and then transferred that function on a contract basis to an external party, this represents a shift in labour and economic terms from manufacturing to professional cleaning services. However, from the climate perspective, the development of such arrangements provides interesting possibilities to optimizing production. For instance, a company that specializes in cleaning may use specialist equipment that is not economical when bought for only one operation. More development may also be expected in cleaning technology since cleaning is the core business of the service company. Similarly, outsourcing of transport and logistics may lead to a reduced need for warehouses and optimization of materials and energy flows. Such specialist parties have an economic interest in reducing materials and energy flows, and their emergence may produce new actors that develop such applications with positive effects on the climate.

An example is Diverseylever, a Unilever subsidiary that provides cleaning materials to business markets. For industrial washing processes, it also offers maintenance and advisory services based on its proprietary automated dispensing and monitoring systems. Computers are used as a tool for providing the service of product extension. The environmental benefits are reduced resource consumption. The economic benefits are reduced costs and higher washing quality.

Positive effects of product-service systems on material use

In general, the shift to services is expected to have significant consequences for the use of materials. Some authors believe that this concept has the potential to reduce environmental impact by a factor of 4 to 20 (Weizsacker et al., 1998; Roy, 2000). This is due to the fact that adding a service component to products may lead to more efficient product use and improved life-cycle management.

Emission reduction due to substitution of products by services is also demonstrated by de Vries et al. (2000), which describes a scenario-based model analysis of energy and material use at the macro level. In the scenario, a significant reduction of greenhouse gas emissions is demonstrated, based on global cooperation in technology transfer and a shift to a service-based economy.

Besides model results, actual trends are also visible in the national economies of Europe and North America. There are several indicators that point to a dramatic shift in employment from the agricultural and industrial sector to the service sector. Furthermore, the economic value of the service sector is increasing. Some expect the share of industrial production to decline to levels as low as those of the agricultural sector half a century ago. Consequently, the amount of energy and materials necessary for a particular value of economic production is also bound to be reduced.

Rebound effects

Besides the positive effects of product-service systems, there are also several rebound effects that may reduce the environmental benefits of these systems.

The first rebound effect is based on the marketing and economic sub-disciplines that deal with product-service combinations. The literature on this devotes a lot of attention to the dynamics of product-service systems, which means that products and services are increasingly interconnected (Oppedijk and van Veen, 1999). This literature states that services are added to products to satisfy consumer needs and are actually strategies to increase sales and stabilize market shares. Therefore, services stimulate economic growth. In addition, product-service systems consist of material as well as service elements. On the basis of these studies, the emergence of 'pure' services is not to be expected. Thus, according to this literature, services will only lead to more consumption and most likely also to more material use.

Another rebound effect may be the following. When regarded from the viewpoint of the consumer, product-service systems are attractive since they may reduce the amount of time consumers have to spend in fulfilling their needs. This may be attractive for the consumer but not for the climate problem if this extra time is used for energy-intensive activities. For instance, a cleaning and washing service might deliver first-order savings in materials and energy (through sharing of machines, for example), but the time saved might be spent on increased travel or consumption of other products requiring additional materials and energy.

Thus, services may lead to more material-efficient fulfilment of human needs but two rebound effects (increased consumption and increased spare time) are likely to reduce the environmental benefits. We therefore expect a weak, not a strong, dematerialization effect from these new concepts. We expect the latter will only be possible if major changes occur in attitudes to consumption. However, a lot more research is needed to really understand the effect of service initiatives on consumption.

That brings us to the last uncertainty about the potential effects of product-service systems on CO_2 emissions. It turns out that the optimism about services is based on a limited number of promising examples, all of which show dematerialization effects. However, it is not known what can be expected of these developments for the total economy. In addition, the effects of the rapid rise of the ICT sector, which makes up an increasing part of the total economy, are only estimated and not calculated. It turns out that the statistical basis for studying the impact of these new developments is very poor. In national statistics, the 'service industries' category changes very rapidly. The composition of the service sector five years ago is completely different from what it is in the most recent figures. To find out more about the influence of service activities on material use and the reduction of CO_2 emissions, much more detailed statistical data are necessary. Better knowledge about the impact of services is an absolute necessity before successful policies in this field can be designed.

Environmental policy to optimize material chains

The preceding sections showed that many measures are available to improve the efficiency of material use in economies. We have also shown that focusing on more efficient material management is likely to be an effective instrument to reduce greenhouse gas emissions. Some of the measures are already part of national policy plans, like material recycling and environmentally friendly product design. However, material policies are very seldom designed from the viewpoint of reducing greenhouse gas emissions. That would be trend-breaking in itself.

According to the classification presented in this chapter, policy efforts could be targeted at the following categories of material efficiency:

1. technical measures in individual life-cycle stages
2. management of the entire material-product chain
3. sustainable product-service systems.

In regard to point 1, technical measures in individual life-cycle stages could well lead to large reductions in CO_2 emissions. Earlier studies on packaging have shown that simple technical measures (e.g., lighter packaging) could reduce the related greenhouse gas emissions by 20 per cent (Hekkert et al., 2000a). The model described above also showed that considerable gains could still be made in optimizing the waste-management stage. In general, these measures are not very difficult for governments to stimulate. A lot of experience has already been gained in the area of energy efficiency. Exactly the same efforts can be made to reduce material demand.

Stimulating measures in the second category is much more difficult. Significant reductions in greenhouse gas emissions could be reached by improved management of entire material-product chains. Efficient chain management requires that the different actors in a particular chain work together to reach a desired situation. This calls for all actors in the chain to be aware of their responsibilities towards improving the efficiency of the chain. It means that actors must learn to understand the specific stakes that are involved for the other actors in the chain. Reaching these goals is likely to require frequent meetings. Getting the different actors to work together is unlikely to be an autonomous development. Governments can initiate this process by taking the lead in organizing these meetings or by designing certain chain-management regulations in order to force the actors in the entire chain to work together in order to reach policy goals.

In The Netherlands, initial efforts have been made in this area. Efficient chain management to reduce greenhouse gas emissions has been made part of voluntary agreements on energy efficiency with industrial parties. In other words, industries are allowed to meet their energy-efficiency targets in part by improving material efficiency. Since these measures are often cheaper than energy-efficiency measures (where the cheapest measures have already been taken in The Netherlands), the industrial parties are stimulated to invite other actors involved in the same material chain to reach efficiency goals together. The different parties are encouraged to come up with a fair system for dividing the credits for reducing CO_2 emissions among the different actors involved.

Other measures that may improve life-cycle efficiency of materials would be stimulating chain responsibility instead of product responsibility, setting strict standards for recycled content in products and reusability of products, setting up systems for taking back used products in order to enhance recycling rates and stimulating repair facilities to extend the useful life of products.

When policies are developed to optimize material life-cycle management, one should bear in mind the fact that many measures already taken in this area are focused on different goals, such as reducing waste (waste policies) or reducing the environmental impact of product manufacturing (product policies). Specific attention to material life cycles with low greenhouse gas emissions should be integrated with other policies that currently affect material life cycles.

On the third point, policies could also focus on the shift from producing and consuming products towards producing and consuming services. It is not yet clear how much greenhouse gas emissions can be reduced with this type of policy. Many of the mechanisms behind the shift from products to services are also unclear. In fact, there have been only a few successful case studies, and it is difficult to generalize from them. Therefore, it can be said that we do not know a great deal about measures of this type. The most fruitful policy actions in this category of improvement are therefore the development of a wide range of pilot projects to learn more about the dynamics of these measures. Insight into potential reductions of greenhouse gas emissions, the potential market penetration and potential rebound effects is sorely needed.

Conclusions

History shows that modern economies have great difficulty achieving strong dematerialization, which means an absolute reduction in material use in a growing economy. Far more policy-related attention should be paid to more efficient material use in order to reach strong dematerialization and thereby reduce greenhouse gas emissions. There are two reasons for this. First, the use of materials has a significant effect on the emissions of energy-related greenhouse gases, and second, there is considerable potential for a large reduction of greenhouse gas emissions in more efficient material management. More efficient material management can be reached in each individual stage of the material life cycle, by optimizing the total life cycle and by stimulating a shift from consuming products to consuming services.

The two visions on the future energy system both include improved material management as an important tool to reduce greenhouse gas emissions. This is justifiable in light of the conclusions of this chapter. However, one should bear in mind that serious policy efforts are needed to reach this desired situation.

References

Bouwman, M.E., and H.C. Moll, *Status Quo and Expectations concerning the material composition of road vehicles and consequences for energy use*, Report number 91, IVEM, Groningen, 1997.

Cleveland, C. J. and M. Ruth, Indicators of Dematerialization and the Materials Intensity of Use. *Journal of Industrial Ecology* 2(3), 1999, pp.15–50.

Cochran, C. N., Long-Term Substitution Dynamics of Basic Materials in Manufacture. *Materials and Society* 12(2), 1988, pp.125–150.

Daniels, B. and H. C. Moll, *The Base Metal Industry: Technological Descriptions of Processes and Production Routes*. Groningen, Rijksuniversiteit Groningen - IVEM, 1997.

de Bruyn, S. M. and J. B. Opschoor, Developments in the Throughput-Income Relationship: Theoretical and Empirical Observations. *Ecological Economics* (20), 1997, pp.255–268.

de Vries, B., J. Bollen, L. Bouwman, M. den Elzen, M. A. Janssen, E. Kreileman, and R. Leemans, Greenhouse Gas Emissions in an Equity-, Environment- and Service-Oriented World: An IMAGE-Based Scenario for the Next Century. *Technological Forecasting and Social Change* 63, 2000, pp. 137–174.

Duin, R. v., M. P. Hekkert, J. de Beer, and M. Kerssemeeckers, *Inventarisatie en Classificatie van EZP opties*. Ernst, Bureau B&G, 1999.

EPA, *Greenhouse Gas Emissions From Management of Selected Materials in Municipal Solid Waste*. United States Environmental Protection Agency, Washington D.C, 1998.

Gielen, D.J., *Materialising Dematerialization. Integrated energy and materials systems engineering for greenhouse gas mitigation*. Delft. Ph.D. Thesis, Technical University Delft, 1999.

Goverse, T., M. P. Hekkert, P. Groenewegen, E. Worrell, and R. E. H. M. Smits, Wood in the Residential Sector: Opportunities and Constraints. *Resources, Conservation and Recycling* 34, 2001, pp. 53-74.

Fraanje, P. J., Cascading of Renewable Resources: Hemp and Reed. *Industrial Crops and Products* 6, 1997, pp. 201–212.

Hekkert, M. P., *Improving Material Management to Reduce Greenhouse Gas Emissions*. Ph.D. thesis. Utrecht, Utrecht University, 2000.

Hekkert, M.P. and E. Worrell, *Technology Characterization of Natural Organic Materials*, Department of Science, Technology and Society, Utrecht University, Utrecht, 1999.

Hekkert, M. P., L. A. J. Joosten, and E. Worrell, Reduction of CO_2 Emissions by Improved Management of Material and Product Use: The Case of Transport Packaging. *Resources, Conservation and Recycling* 30, 2000a, pp. 1–27.

Hekkert, M. P., L. A. J. Joosten, E. Worrell, and W. C. Turkenburg, Reduction of CO_2 Emissions by Improved Management of Material and Product Use: The Case of Primary Packaging, *Resources, Conservation and Recycling* 29, 2000b, pp. 33–64.

Hekkert, M. P., D. J. Gielen, E. Worrell, and W. C. Turkenburg, Wrapping up Greenhouse Gas Emissions. *Journal of Industrial Ecology* 5 (1), 2002.

Jänicke, M., M. Binder, and H. Mönch, Dirty Industries: Patterns of Change in Industrial Countries. *Environmental and Resource Economics* 9, 1997, 467–491.

Jänicke, M., H. Monch, T. Ranneberg, and U. E. Simonis, Structural Change and Environmental Impact: Empirical Evidence on 31 Countries in East and West. *Environmental Monitoring and Assessment* 12, 1989, pp. 99–114.

Joosten, L. A. J., *Process Data Descriptions for the Production of Synthetic Organic Materials*. Utrecht, Utrecht University, Department of Science, Technology, and Society, 1998.

Malenbaum, W., *World Demand for Raw Material in 1985 and 2000*. New York, McGraw-Hill, Inc.,1978.

Matos, G. and L. Wagner, Consumption of Materials in the United States, 1900–1995. *Annual Review for Energy and Environment*, 23, 1998, pp. 107–122.

Patel, M., *Closing Carbon Cycles: Carbon Use of Materials in the Context of Resource Efficiency and Climate Change*. Utrecht, Utrecht University, 1999.

Potter, S., R. Roy, and M. Smith, *Exploring Ways to Reach Factor 10+, International Summer Academy on Technology Studies*, 'Strategies of a Sustainable Product Policy', Deutschlandberg, Austria 9–15 July 2000.

Price, L., L. Michaelis, E. Worrell, and M. Khrushch, Sectoral Trends and Driving Forces of Global Energy Use and Greenhouse Gas Emissions. *Mitigation and Adaption Strategies for Global Change* 3, 1998, pp. 263–391.

Roberts, M. C., What Caused the Slack Demand for Metals after 1974?, *Resources Policy* 14, 1988, pp. 231–246.

Roberts, M. C., Predicting Metal Consumption, the Case of US Steel. *Resources Policy* 16, 1990, pp. 15–50.

Roy, R., Sustainable Product-Service Systems. *Futures* 32, 2000, pp. 289–299.

Schmidt-Bleek, F., *Wieviel Umwelt Braucht die Mensch. Mips Das Mass für Ökologische Wirtschaft*. Berlin, Birkhäuser, 1994.

Weizsacker, E. v., A. B. Lovins, and L. H. Lovins, *Factor Four, Doubling Wealth, Halving Resource Use*. London, Earthscan Publications Limited, 1998.

Wernick, I. K., R. Herman, S. Govind, and J. H. Ausubel, Materialization and Dematerialization: Measures and Trends. *Daedalus* 125, 1995, pp. 171–198.

Williams, R. H., E. D. Larson, and M. H. Ross, Materials, Affluence, and Industrial Energy Use. *Annual Review of Energy and Environment* 12, 1987, pp.99–144.

Worrell, E., B. Meuleman, and K. Blok, Energy Savings by Efficient Application of Fertilizer. *Resources, Conservation and Recycling* 13, 1995a, pp. 233–250.

Worrell, E., A. P. C. Faaij, G. J. M. Phylipsen, and K. Blok, An Approach for Analysing the Potential for Material Efficiency Improvement. *Resources, Conservation and Recycling*, 1995b, pp. 215–232.

Chapter 8

The contribution of Information and Communication Technologies to the transition towards a climate-neutral society

Adriaan Slob and Marc van Lieshout

In this chapter...

... after having looked at material efficiency as a significant option for reducing greenhouse gases, we will now focus on another important, evolving area of technology: information and communication technology (ICT). Obviously, information and communication technologies have changed the world enormously in the last decade. Some authors refer to ICT as a major resource for increasing the eco-efficiency of many activities. But at the same time it is clear that ICT has also stimulated economic growth and its consequent impact on energy use. This chapter assesses the contribution that ICT could make to realizing a climate-neutral society. To assess the impact of this new technology, ICT will be placed in its social context of behavioural and institutional arrangements. This chapter will show that ICT may have some benefits but that the potential savings are modest, and rebound effects will need to be addressed as well.

Introduction

The rapid pace of the development of Information and Communication Technology (ICT) could have a major impact on our economy and thus on energy use. In this chapter we will elaborate on this phenomenon and try to assess what kind of impact could result for the Dutch and global situation. ICT fits in both visions that were presented in Chapter 2, with small differences. In vision A the use of ICT is widespread, sustaining a global market. In vision B ICT is especially used for the intensive and efficient use of space and for efficiency of mobility and transport. Although technologies, like ICT, can change quite quickly, our behaviour and institutions will not change that fast. Therefore, we will argue that ICT is not a 'miracle technology' changing our behaviour and habits, but should rather be considered as a tool that sustains and enforces our existing institutions and behaviour. ICT can have both negative and positive impacts for climate change. We expect that without changes in our behaviour, institutional arrangements and socio-technical system, ICT will not have very obvious effects for climate change. We expect that ICT applications will result in a small reduction of CO_2 emissions, far less than the required 80 per cent that was mentioned in Chapter 2.

The Netherlands Bureau for Economic Policy Analysis (CPB) (CPB, 2000) suggests that ICT has the character of a 'breakthrough technology', similar to the invention of the steam engine and electricity. ICT could have a major impact on how we work, live, shop, spend our leisure time, move around, etc. ICT applications make it possible to perform certain actions faster, in a smarter way and at any time and anywhere. Such changes will be felt at the micro-level (the behaviour of consumers and producers), but also at macro-level (the economy, globalization).

In this chapter we focus on the implications for energy consumption on different scales (global, national and sector-specific). The development of ICT can be looked at in a number of totally different ways. We have opted for an approach in which we consider the changes that technology causes in a technological *system* in a *social context*. The role of the government, the market, companies and consumers are assessed in a balanced way. So we consider not only the technology itself (in other words, we do not regard the development of technology as autonomous), but review developments at system level, in which different related technologies, infrastructure, and last but not least various societal actors play an important role. The use of technology by the various actors, how it is applied, will have an effect on energy consumption. In the next section we will start by defining what we mean by technology in the social context. Then we look at the technological developments surrounding ICT and their social implications. In the following section we look at the relationship between these developments and climate change and we discuss the possible effects and the conditions under which these effects could occur. Finally we will present the main conclusions and research questions.

Technology in a social context

To look at the entire technological system in the social context, we use a simple but transparent model (Slob, 1999) that has three central elements:

- **Arrangements**: all institutional, organizational or commercial arrangements that can guide technological development and the behaviour of individuals and companies in a particular direction. These include environmental laws and regulations, but also tax measures or public information, the institutions to implement them, etc. It also includes services as a commercial arrangement whereby technology is used more efficiently and at the same time behaviour is influenced.
- **Behaviour**: this refers to the behaviour of companies and individuals. The behaviour in the technological system will have a major impact on the resulting environmental effects. Both quantitative and qualitative aspects are relevant in this regard. The quantitative aspects (volume) of production and consumption are closely related to the quantity (amount) of the environmental aspects, while the qualitative aspects (quality of products and services) are largely responsible for the nature of the environmental aspects.
- **Technology**: the artefacts that are developed and used to meet the needs of consumers and producers. In this context, improved and innovative technologies that cause far less pollution are interesting. These technologies are also referred to with the 'factor-metaphor', which describes the extent to which the technology is improved (e.g. by a factor of 4, a factor of 20).

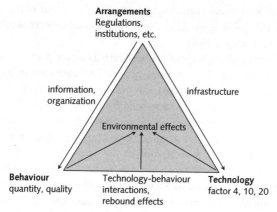

Figure 8.1 *Technology in social context*

These three elements are not independent but interact along the sides of the triangle:

- **Organization and information** as the link between arrangements and behaviour. Information and organization can be seen as a weak form of guiding behaviour. Environmental information and eco teams are examples of this. On the other hand, information and organization can be used to provide feedback to arrangements (monitoring): have the envisaged effects of policy or regulation been achieved? Should targets be revised?

- **Infrastructure** as a physical form of arrangement, particularly with a view to directing the application of technology. We also include the R&D infrastructure which directs the development of technology in this. Infrastructure offers an important (technological) context for the application of technology and can therefore direct the efficient use of technology and the environmental benefits arising from it to a large extent. Infrastructure is essential if we are to profit from new technological possibilities. After all, new technological possibilities sometimes require an entirely new infrastructure. At the same time, the old infrastructure can sometimes slow the process enormously. There may be interests associated with the old infrastructure (especially those of the owner of the infrastructure) which can form a significant barrier to the optimal use or introduction of new infrastructures. The discussion in The Netherlands in the 1980s concerning the delivery of electricity from windmills to the national grid is an example of this. This only changed when the Electricity Act was amended (an arrangement).

- **Technology-behaviour interactions** determine the way in which the user ultimately deals with the technology. There is a lot of evidence from the past showing the existence of a rebound effect: the phenomenon that the potential environmental benefits resulting from technological improvements are not secured because behaviour also changes (see Slob et al., 1996). The technology is used differently than was intended or new applications are thought up so that the environmental effects arising from the volume of use cancel out the environmental consequences of more efficient use. Examples of this are the introduction of the long-life light bulb, which led to new applications (outdoor lighting etc.) rather than the desired substitution of ordinary light bulbs in the living room. The behaviour of the user has to be carefully considered in estimating the environmental aspects resulting from the introduction of new technology.

The behaviour of users and the technology used (artefacts) are jointly responsible for the resulting effects on the environment. The arrangements influence environmental effects via the technology used and the manner in which it is used (the behaviour). New technology induces new behaviour: new, previously unexplored applications (behaviour options) are found for new technologies. Changes in behaviour can in turn call for new arrangements. Infrastructures can also induce certain behaviour, while at the same time old infrastructures can prolong certain forms of behaviour. The introduction of a new technology leads to change at each corner of the triangle. So when considering new technology, one also has to examine the other aspects of the triangle: behaviour, arrangements, organization, information, infrastructure and the technology-behaviour interactions

ICT developments and their implications

Trends in ICT applications

The pace of innovation in information and communication technologies is high. New generations of products succeed one another in quick succession. For instance, the current expectation is that a mobile telephone will be followed after nine months by the next generation of mobile telephone, which is smaller, uses less energy and provides more services. By 2000, the PC with a 486 processor, which was still 'top of the range' in the beginning of the 1990s, is already 'antique'. The basis for these rapid developments is expressed in Moore's Law (Cohen et al., 2000).

The law, which originates from an observation of Moore in 1965 (Intel, 2001), predicts that the capacity of chips will double every eighteen months while the price remains the same. This law could still apply for many more years (the esti-

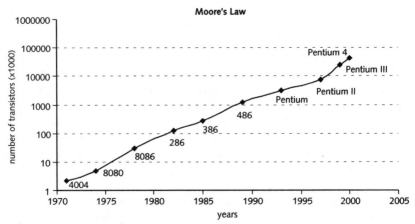

Figure 8.2 *Moore's Law (Intel, 2001)*

mate is until 2020) and forms an important driving force behind numerous ICT trends (miniaturization, etc.).

There is not enough space in this chapter for an in-depth and systematic survey of the developments in the technological basis of ICT. Nor is it crucial for our purpose, since we assume that it is not so much the technology itself as the organization of the technology in specific social arrangements and in specific user contexts that counts. To get a feel for the background to the current developments, we will briefly present a number of interesting trends, which together give an impression of the 'push' side of this wave of technological innovation (see Bozetti, 1999; Bozetti, 2000). We identify five interesting trends:

1. Continuous *miniaturization* of individual components (chips, digital cameras (the size of a sugar cube), mobile telephones, monitors).
2. *Digitization* of text, image, voice and data, making it possible to develop applications that can handle all sorts of data; this enables, for instance, telephony via the Internet (Voice over IP) and, vice versa, Internet via the telephone (WAP: Wireless Application Protocol).
3. *Convergence* of formerly separate networks; telecommunication networks, computer networks and TV networks converging into a single infrastructure carrying text, images, voice and data. Two remarks need to be made here. First, the formerly 'vertical' structure (vertical segmentation: separate telephone, TV and computer networks) is being replaced by a layered structure (horizontal segmentation: infrastructure – network services – applications), which has major repercussions for the organization of market players. Secondly, a single infrastructure that carries everything does not mean that the entire infrastructure is completely transparent; many individual networks are being created.
4. *Integration* of architectures, standards and protocols; the Internet protocol currently seems to be dictating the direction of network developments, from internal business applications to public networks.
5. Added *intelligence* to networks and applications. In a dialogue with the user, Smartbots develop an impression of what the user wants and provide the link between 'front office' and 'back office'.

It is impossible to forecast whether these trends will still manifest themselves in the same way in 50 years' time. At the moment, the technological horizon for most companies extends not much further than two years at most. What we can say with certainty is that for the time being the technical potential of ICT will only increase. The fact that the chip is everywhere nowadays, from refrigerators to high-tech business systems, is known as 'ubiquitous computing' (ICT everywhere, under any circumstances and at any time). Nevertheless, it is going too far to assert that the availability of ICT automatically leads to its application. Against the potential of ICT as an *enabling technology* there is the problem of its integration and acceptance in society. As emerged in the introduction, we assume that ICT and the social context within which it has to function go hand in hand. Each influences the other and they cannot be seen separately from each other. There are plenty of examples of promising technologies that failed to take off or whose use fell far short of expectations. While we can regard the rapid introduction of the mobile telephone as a success story, the other side of the coin is represented by the failed introduction of the electronic purse (known in The Netherlands as the Chipknip and Chipper), the failure of digital money to get off the ground and the failure of steadily more advanced office accessories to increase productivity by as much as anticipated (CPB, 2000). So the success stories have to be treated with a certain caution.

The new economy

Many scientists believe that ICT is turning old economic laws on their head. According to them, a transition is currently taking place from a product economy, in which supply and demand focus on finding a market equilibrium for scarce goods, to a service economy, where the same service can be repeatedly sold at marginal additional cost. Illustrative of the current economic order is the fact that mobile telephones are given away or sold for a very low price with a subscription. The profit is no longer to be found in the added value of the physical product but in the services provided along with the telephone (the use of the telephone, the SMS options, its use abroad, connection to the Internet). The new economy is further characterized by the emergence of new giants in the telecommunication, media and amusement markets. Enormous sums are being paid in the mega-mergers between formerly independent companies (media giant Time Warner merges its activities with Internet behemoth America Online, and amusement colossus Endemol goes into partnership with Spain's telecommunication company Telefonica). The proceeds from the auction of the third generation of mobile telephone frequencies (UMTS) were enormous in some countries, like Germany and the UK, while in other countries, such as The Netherlands, the auction wasn't regarded a big success. The sale of licences raised 80 billion guilders in the United Kingdom, and as much as 120 billion guilders in Germany. The sale boosted economic growth in Germany by a few tenths of a percentage point in 2000 and one newspaper reported that the German government would as a result close the financial year with a budget surplus rather than a deficit (*De Volkskrant*, 30 August 2000). One final factor that we want to point out is the imbalance between the 'virtual' value, particularly of Internet companies, and their actual earnings. The share value of large new e-services, like the famous electronic bookstore Amazon.com, the browser Netscape and many other familiar and less familiar initiatives, is far higher than the sales of those companies would justify. In fact, the developments in the second half of 2001 show that the rules of the 'new economy' are severely challenged by the rules of the 'old economy'. Incurring high debt can cause severe problems. The former Dutch monopolist in telecommunications is in serious financial trouble due to its high investments in UMTS and in foreign partners. In the fall of 2001 the biggest Dutch micro-electronics firm, Philips, requested shorter working time for most of its employees. The hardware industry and the information service industry are experiencing a severe fall-off in demand for chips and new services and consultancy. Analysts proclaim the fall-off to be part of a shake-out of the market and expect new growth figures for the ICT business in the second half of 2002.

Notwithstanding this temporary drawback, the ICT sector continues to grow in importance. According to a recently published study by the CPB (2000), the ICT sector makes a relatively large contribution to economic growth. For the first time the increase in labour productivity in the ICT sector is also higher than the average. The CPB study breaks the ICT sector down into producers of ICT products and ICT service providers. The latter category encompasses services relating to ICT products (system management, software maintenance, installation of software) and telecommunication facilities. In relation to the environment, it is interesting to note that, according to some, for the first time economic growth and environmental burden have been decoupled. A study in the US shows that the economic growth in the last two years has not led to a proportionate increase in the environmental burden (Romm, 1999). This is partly attributed to the new economy, in which services cause less environmental harm than the traditional economy with its material production. However, the arguments presented in support of the correlation shown in the study are not particularly strong.

ICT and sustainability

ICT can contribute to reducing environmental effects in different ways via different mechanisms (see for example Ducatel et al., 1999).

First, the trend towards miniaturization and digitization can contribute to *dematerialization* of the economy. After all, miniaturization means that in computers and similar products fewer materials are needed to perform the same function. Digitization means that products are available digitally rather than materially, such as sending an e-mail instead of a letter.

Second, miniaturization means that ICT can be built into a growing number of appliances, so that more and more *smarter, intelligent appliances* will come on to the market. This will make it possible for appliances to provide instructions for use (feedback technology) or programme themselves for optimal use. In other words, it will be possible to delegate a user's behaviour in an intelligent manner to appliances.

Third, there can be a *decoupling of time and place*. Thanks to the increased possibilities of communication, actions can be carried out from anywhere. This means that various activities can be performed from the home: shopping, learning, working, etc. The potential effect of this for the environment will be substantial if there is *substitution* of movements. We will look at this in more detail in the following section.

Fourth, ICT can lead to *more efficient organization* which is facilitated by the faster exchange of information in companies and in chains. Various activities can be performed more efficiently.

Fifth, it is increasingly possible to provide consumers with *tailor-made information* for purchasing decisions and about the use of appliances. Moreover, it may be possible to buy environmentally-friendly products via the Internet. Through its 'catch-all effect' ICT solves the problem of a diffuse market in which the buyers are initially few and far between.

But the relationship between ICT and sustainability is double-edged: substitution (of raw materials, transport movements) and optimization lead on the one hand to a reduction in the use of raw materials, but on the other hand these gains are totally nullified by the explosive growth in the use of ICT. The few thousand mainframes of 25 years ago bear no relation to the many tens to hundreds of millions of computers worldwide with a shelf-life of little more than three years. The rise of e-mail has up to now led to scarcely any reduction in the flow of paper. The arrival of Internet companies in Amsterdam has led to extra demand for electricity. A recent report (Huiberts, 2001) estimated the electricity use of the ICT sector in The Netherlands at more than 500 GWh, which is about 0.5 per cent of total Dutch electricity use. This electricity use is expected to grow to about 6700 GWh (about 6 per cent of total Dutch electricity use) in 2005 and to about 11,100 GWh (about 10 per cent of total Dutch electricity use) in 2009. The electricity use of the ICT sector is thus expected to grow enormously. However, there are also positive reports (Forsebäck, 2000). The Japanese Telecommunication Council forecasts a reduction of 3.81 MTon of CO_2 by 2008 (or approximately 7 per cent of Japan's CO_2 emissions) if teleworking, intelligent transport systems, paper reduction through the use of Local Area Networks, household management systems, electronic publishing and e-learning are exploited in a realistic fashion. The ultimate question is whether the balance of savings and new and additional activities is still positive. The uncertainties in this respect are considerable.

ICT and climate change

In this section we look at the possible implications of the implementation of ICT for energy consumption. In Chapter 2 it is argued that an 80 per cent reduction of greenhouse gas emissions is required in 2050 compared to 1990 levels. In this section we will assess how ICT can contribute to this reduction by analysing the possible effects of ICT in the various sectors of the economy. The context of the use of ICT in vision A (see Chapter 2) is the 'internationally oriented Global Village'. In this scenario, therefore, ICT use is concentrated on e-commerce and on efficiency of material flows in the global economic system. In vision B the context of ICT use is the regional economic and social system. One can expect the use of ICT in this scenario to be directed more to the efficient use of space and to mobility and transport efficiency.

The home and the built-up environment

The home environment will become more important in the society of the future. Besides being a place to live, many activities in the house lend themselves to digitization and development of virtual services, ranging from television viewing and telephoning to electronic banking, telelearning, teleshopping and teleworking.[1] The home is an ideal place for decoupling processes formerly linked in time and space. A person can bank electronically at any time of the day and no longer needs to visit the bank.[2] The energy saving from teleactivities arises from the fact that the suppliers (for example, banks) need fewer people, fewer offices, fewer activities (in fact the work shifts to the consumer), that less transport and distribution of physical goods is needed (video-on-line means a reduction of transport between the

---

[1] We return to teleworking in the section on transport.
[2] As Forsebäck argued in a recent study, we need banking but we don't need banks (Forsebäck, 2000).

video store and the central warehouses) and the saving of paper (thanks to electronic communication). These savings are empirically demonstrable but are probably not significant in terms of the total energy consumption of a household.[3]

A second aspect is the degree to which ICT regulates overall energy consumption, starting with the provision of information prior to the planned purchase to the installation of equipment or software in the information architecture in the home and optimization of the use in combination with the intelligence built in to appliances. Even energy management itself has energy-saving potential. But the direct contribution of ICT to the savings will be modest, and, because more appliances will be purchased with a higher replacement rate, may even be nil or negative. If the new style of living also involves the need for more rooms (for example because of teleworking, but also in connection with higher demands on home comfort) this will have a negative effect on the amount of energy needed. It should also be pointed out that although appliances may become more energy efficient, the energy balance will shift more towards production, distribution and disposal of the appliances.

A final aspect that we want to mention in connection with homes is the way in which they are an element of spatial planning. If ICT contributes to reducing the importance of distance as a factor in the choice of where one lives, a possible consequence is that distances travelled (to work, to the shopping centre, to leisure centres, to school) will increase. If this is attributed to a household, it generates additional energy consumption. On the other hand, it is conceivable that carefully considered locations for new housing in relation to each other and to business parks, shopping centres, schools, etc. could lead to energy saving. It is also possible that advantages of scale could arise in the intelligent linking of energy infrastructures between home and working environments (combined heat and power generation plants in districts, district heating complexes, options for supplying electricity back to the national grid, etc.). Up to now, there has been little or no research into such possibilities and effects of spatial planning. Generally speaking, it is disappointing to have to observe that ICT developments are very seldom assessed for their spatial planning implications, and vice versa. ICT was scarcely mentioned in the debate that preceded the Dutch Fifth Policy Memorandum on Spatial Planning. In a project like Gigaport, the aim of which is to produce the successor to the existing Internet infrastructure and applications that can use it, considerations of sustainability play scarcely any role.[4]

---

[3] In a case study of the replacement of individual answering machines with a single central *on-line* answering service, Telia produced energy savings of more than a factor 200. Although when measured over a million households it is a substantial effect, the benefits per household remain small.
[4] Interview with Ms. J. Tammenoms-Bakker, April 2000.

Transport and mobility

The use of ICT can promote sustainability in the traffic and transport sector in a number of ways:

1. by optimizing the use of the infrastructure (including promotion of the 'modal split');
2. by optimizing the logistics chain;
3. by reducing the pressure on mobility through alternatives (teleworking, teleshopping, telelearning);
4. with intelligent vehicles and monitoring of cargoes.

The optimization of the infrastructure does not automatically lead to a reduction of CO_2 emissions. More efficient use of the infrastructure could help to spread traffic through the day, so that if the traffic flow increases there will be greater emissions of CO_2. Promoting the modal split is a way of stimulating the most efficient form of transport from an energy perspective (especially transport by water compared with transport by road). However, ICT mainly further reinforces the competitive advantage of the door-to-door approach of road transport, in part because the use of ICT increases the opportunities for customized service.

Optimizing the logistics chain fits in with the *Just in Time* approach. Greater control of the entire transport and distribution process means that smaller inventories need to be held, and consequently smaller storage systems are needed. With the use of a comprehensive logistics system, the utilization of capacity of goods transport can be increased: fewer empty runs and optimization of the length of trips. But in a recent study Transport en Logistiek Nederland (TLN) pointed to the likely increase in the number of trips with (small) vans if e-commerce really takes off (TLN, 2000). In the next five years TLN expects an increase of 38 per cent in the number of trips for freight transport. Of this 38 per cent increase, 21 per cent will be 'autonomous' growth and 17 per cent due to developments in e-commerce (8 per cent growth in the Business-to-consumer sector, 9 per cent in the Business-to-Business sector). A scenario study by the ministries of Transport, Public Works and Water Management, Economic Affairs and Housing, Spatial Planning and the Environment concluded that with the conscious use of ICT tools CO_2 emissions could decline sharply but would probably still remain above the target of 168 billion kilograms (Nederland Digitaal, 2000).

Teleworking leads to substitution of commuter traffic. The savings could rise to 40 per cent of the number of trips per week (ICT & Sustainability, 2000; Forsebäck, 2000, p34). The increased use due to the car becoming available generally seems to be less than the savings (Puylaert et al., 1999). In other words, the sum of the effect of a car being freed up for secondary purposes (social and recreational traffic), other work-related movements or instigation of new commuter traffic is less than the savings achieved. In combination with the assumption that the house is a more energy-efficient place to work than the office (ICT & Sustainability, 2000), large-scale introduction of teleworking could make a positive contribution to reducing CO_2 emissions. Besides teleworking, teleshopping can also save trips. Up to now the expectations for teleshopping have scarcely been converted into quantifiable data. Following the TLN study, a certain degree of scepticism is therefore justified about the direction these developments will take.

Industry

There is some literature on the effects of ICT at macro-level, but we found no literature for specific sectors of industry. What follows, therefore, is based mainly on our own forecasts of those effects. ICT could have two major influences on industry. On the one hand, optimization in companies and collaboration in chains so that energy and materials can be used more efficiently. On the other hand, ICT could reinforce the existing trend of the growing importance of services in the economy, because services can be provided via e-commerce and tailored to the individual customer. Transport activities relating to industry are not considered here. They were discussed in the previous section.

The greater accessibility of information and the speed with which information is exchanged will make it easier for companies to share information worldwide about the environment and to monitor flows of raw materials and waste substances. This information is essential for chain management, where the various steps in the production process, from raw material to product to waste, can be linked in such a way as to keep pollution to a minimum. In this respect, ICT acts as a facilitator for collaboration between suppliers and customers and hence can help to optimize the entire chain. This aspect will have a positive effect, particularly for the substances emitted to the environment and not so much for energy consumption. To grasp these benefits an organizational structure setting out the protocols for information exchange (method of exchanging information, infrastructure, assurance of reliability, confidentiality, etc.) is needed.

Various industrial production processes can be made more intelligent and can be better monitored with ICT applications. This could generate savings of raw materials and energy, although for the time being we do not estimate this effect as being very great. Production processes in companies are already largely optimized. ICT could make a minor contribution to this.

Finally, ICT can help to increase the transparency of the market for raw materials, waste and residual products. For instance, industrial ecological complexes could be formed at local level, in which the waste from one company becomes the raw material for another, or in which companies swap heat and electricity with each other. But ICT can contribute little to the creation of these complexes: it is not so much the 'market' that is the problem but rather the need for collaboration between companies, which poses a number of dilemmas, such as the creation of mutual dependency, the division of costs and benefits and the risks for business operations. The introduction of ICT is expected to have at best a modest impact on energy consumption.

The trend towards the increasing importance of services in the economy has been apparent for years. An important question, therefore, is how the industrial sector will develop in relation to the service sector. Will we have a full service economy in 2050 or will there still be substantial industrial activity?

From the ICT perspective, what we can say is that ICT can make a significant contribution to the creation, sale and delivery of services. The literature (Bilderbeek et al., 1998; Miles et al., 1999) shows that ICT plays an important role in service innovation. Many new services are born from ICT applications. Take for instance the 'booming business' of e-commerce. It is therefore likely that ICT will reinforce the trend towards the service economy.

What does this mean for the likely environmental effects? Generally speaking, environmental scientists take a very positive attitude towards services, based on the idea that the material component in services is smaller than in products. However, in another study we showed that this relationship is not so straightforward (Nijhuis et al., 2000). After all, services often also need products (for example, computers when we are talking about e-commerce), and since people have to do the work there is often a significant transport component involved in getting the service providers to the client.

Another aspect that emerged in this study was that the 'rebound effect' of services can be considerable. Simple access to the purchase of services can easily lead to far more services being bought than would be necessary purely on the basis of substitution. The volume of demand hence increases, as do the related environmental aspects, so savings can be cancelled out. Chapter 6 also refers to this phenomenon.

We therefore do not simply assume that an increasingly service-oriented economy will automatically lead to reductions of emissions or energy saving.

Agriculture and food

We will look here only at ICT applications in the production phase and not in the processing or consumption phase. In the previous section we considered the consequences for industry and there is no reason to believe the mechanisms described above should work differently for the food industry. At the same time, we considered the consumption phase in the section on 'home'.

Precision agriculture is a promising application of ICT for the cultivation of crops in agriculture. Precision agriculture is a form of farming where growers respond to the natural variability of the soil with the help of different technologies. By sowing, tackling weeds and diseases, applying nitrogen and harvesting according to the requirements of a specific location, it is possible to improve the quality of the harvest, increase yields and curb harmful effects for the environment.

Satellites are an important element of precision agriculture, as with the aid of GPS (Global Positioning Systems) objects, in this case harvesters and sowing machines, can be followed and steered. The yield from a plot is measured using a harvesting machine fitted with a yield meter and a GPS. The result is a yield map, which shows the yields for each location in the plot of land. The differences in yields are explained, for instance by differences in soil composition, ground water level, etc. The data acquired in this way are processed and converted into charts for manure spreading, spraying and sowing. With the aid of sensor technology, muck spreaders, sprays for pesticides and sowing machines can adapt the doses to the exact needs of a particular point in the plot. All the calculations are entered in a single central computer, which helps the farmer to determine the optimal time and route for harvesting. Precision agriculture increases productivity, improves the quality of the crops and the efficiency and effectiveness of the use of resources such as seed, manure and pesticides (Nijhuis, 1999; Nijhuis et al., 2000). Precision agriculture therefore contributes mainly to the efficient use of raw materials and can also help with the more efficient use of energy. We do not feel the effect on energy consumption will be particularly large.

Processes in the horticulture and livestock sectors are already heavily mechanized. ICT applications could make these processes even smarter and further optimize them. We do not estimate the effect on energy consumption as particularly great.

The effect of ICT on climate change

In this section we will try to translate the applications of ICT technology described in the previous section into the consequences for energy consumption in the various sectors. We wish to stress that these estimates are very tentative. We could not find quantitative data in the literature, nor is this the time or place for a more indepth analysis. Table 8.1 lists the various applications in the different sectors and explains how these applications contribute to reducing pollution and energy consumption, with references to the different mechanisms mentioned in this chapter (dematerialization, decoupling of time and place, efficient organization, information provision, intelligent appliances).

The table makes a distinction between the *technically feasible* energy savings and the *most likely* energy savings. Under the technically feasible savings we indi-

Table 8.1 *The contribution to energy conservation of the various ICT applications*

	Mechanism[5]	Technically feasible energy saving[6]	Likely energy saving
Home			
- Energy in the home	3, 4, 5	+	+
- Electrical appliances	4, 5	+	0
- On-line services	2	+	0 / –
- Spatial Planning	2, 3, 4	+	0
Transport			
- Infrastructure/ modal split	2, 3	+	–
- Logistics	2, 3, 4, 5	+	+
- Intelligent vehicles	1, 5	+	+
- Teleworking, teleshopping, etc.	2, 3	++	+ / ++
Industry			
- Chain management	3, 4	+	0
- Smart processes	5	+	0
- Services	1, 3	+	–
Agriculture			
- Precision agriculture	1,3,5	+	0

[5] The different mechanisms are: dematerialization (1), decoupling of time and place (2), more efficient organization (3), tailored information (4), intelligent appliances (5).

[6] The estimates are for potential savings that could be realized in the relevant domain by widespread application of ICT. Estimates:
+ = saving of roughly 0 – 20%
++ = saving of roughly 20 – 40%
– = likely rebound effect that will result in extra greenhouse gas emissions

cate the maximum effect that ICT could have if other aspects are ignored, such as behaviour, organization, infrastructure, etc. In the column 'likely energy saving', we indicate what effect is likely if ICT is used in a traditional manner, in other words without a complete 'redesign'.

We estimate that the energy savings actually achieved are likely to be less than what is technically feasible. This relates to the model that we outlined in the beginning of this chapter. In our view, when new technology is introduced, attention should be given to all aspects mentioned there: the behaviour of societal actors, new arrangements, infrastructure, organization and information and the technology-behaviour interactions. Many of the technical possibilities are not properly exploited because of:

- rebound effects: for example, with teleworking the car is not used for work but for extra private trips; a more efficient organization means that there is time left over for extra production, etc. We therefore expect that the extra 'room' created will be swallowed up by new activities, which leads to extra energy consumption;
- too little attention to accompanying organization and infrastructure;
- failure to provide information to the user;
- divided interests and power positions of actors, so that no one takes the lead in directing the use of ICT technology. Neither the government nor companies feel compelled to create new arrangements or institutions.

Table 8.1 shows that our estimates of the potential savings through optimal use of technology are generally modest and nowhere near the 80 per cent reduction that is needed to avoid climate change effects. The savings in transport (traffic and transport) produce the highest returns from a technological perspective. But the interaction between the various aspects is highly complex: optimizing logistics chains, for example, could have a negative effect on the modal split, and to a certain extent also on the use of the infrastructure (more capacity available, hence less need to shift to a different form of transport). Because of the interdependencies between the various actions, the savings cannot simply be added up. From the (we repeat: tentative) table, it follows that we assess the potential savings for the traffic sector as being greater than for the other sectors. The expected savings (from incorporating ICT in current activities, in other words without substantial reversal of trends) in the traffic sector is also the highest. The expected savings in homes, industry and agriculture are slightly positive or possibly even negative.

Since we argued that in vision B ICT will be used to reduce mobility and for transport efficiency, one can expect that in vision B ICT applications will have the biggest impact on reducing greenhouse gas emissions. In contrast, 'rebound effects' are expected to be quite dominant in vision A and therefore one might expect that ICT applications will have a minor effect on greenhouse gas emissions in this vision.

Conclusions about the relations between ICT and climate change

ICT is developing very quickly in a technological sense. But the social context of ICT (behaviour, institutions, arrangements, etc.) is changing relatively slowly. In our view, the social context of ICT is a key factor that dictates the pace of change and determines the way ICT is implemented. Therefore, the social context is a key factor for the implications of ICT developments for climate change. If we want to benefit from possibilities that ICT developments create to mitigate energy use, we have to influence this social context by means of various arrangements and policy options.

This will call for a totally new policy approach that does not exist at this time. This policy approach is connected to the dilemma of a 'small numbers of decision makers' trajectory versus a 'decentralized, multiple and diffuse decisions' trajectory because a lot of societal actors should be involved. It is also connected to the dilemma of 'transition through co-benefits' versus 'inevitability of a special climate change approach' (see last section of Chapter 3). In the case of ICT applications and climate change, we call for a policy approach that is decentralized and diffuse, involves numerous societal actors, is directed towards collaboration by these actors and is attractive to them. This approach directly addresses the dilemma of 'central steering via direct regulation and market incentives' versus 'co-production' (see last section of Chapter 3). It may be clear that in this dilemma we choose a participatory approach that is directed towards involvement of various societal actors (co-production). The issue is too complex to solve in a 'traditional' way with direct regulation and market incentives. In our opinion such a complex issue calls for an approach that is directed towards collective solutions in a multi-relational network of societal actors.

What policy options do we feel can be adopted?

If we want to gain the maximum benefit from the energy-saving effects of ICT applications, one can think of the following measures:

- monitoring the use of ICT in society, which would also have an important function in identifying the following aspects;
- influencing technology-behaviour interactions; preventing rebound effects;
- ensuring that the infrastructure is adequate;
- providing information to users and ensuring there are adequate organizational models for information-exchange and feedback;
- developing intelligent appliances;
- desired rules and regulations.

But who feels compelled to take the initiative to implement the various conceivable and possible arrangements: the government, the private sector? Something will have to be done if we want to avoid a social dilemma where everyone sits and waits for everyone else.

In this context, we call for collaboration in which the public and private sectors jointly study policy options, conduct research, assign responsibilities for actions and carry them out, all with a greater emphasis on shared responsibility and interactive policy making.

The use of ICT *can* contribute to a reduction of the energy use in several sectors, but will never in itself lead to an 80 per cent reduction of greenhouse gas emissions.

Our analysis shows that transport and traffic can make a particularly significant contribution to the reduction of CO_2. Other sectors can also achieve additional energy conservation, but to a lesser extent. If we want to reap these benefits, consideration will have to be given to new institutional arrangements, new organizational and information structures, infrastructure and to understanding and influencing the behaviour of the users of technology. There are enormous uncertainties surrounding the effects of the rapid developments in ICT. The future interaction between the technology and the user (the rebound effect) is particularly difficult to predict. It is precisely this interaction that will account for the ultimate effect on energy consumption. Additional or alternative use of the technology could turn a potential benefit for energy consumption into a disadvantage. Intelligent appliances can be used to direct user behaviour. But there is still little known about technology-behaviour interactions. We call for more research on this subject to increase the understanding of this phenomenon and to come up with suitable policy options to guide the phenomenon.

In this chapter we have placed a question mark against the energy-saving effects of the growing service orientation of the economy. It is impossible to say in advance whether a service will lead to additional energy conservation since services could lead to extra demand as well as extra transport movements. However, services can play a role in saving energy. It is therefore essential to find out more about the relationship between the growth of services and energy consumption and to use the findings to provide more direction for the development of services.

References

Bilderbeek, R, P. den Hertog, G. Marklund and I. Miles, *Services in Innovation: Knowledge Intensive Business Services (KIBS) as co-producers of innovation*, SI4S synthesis paper, stage 3, STEP Group, 1998.

Bozetti, M., *The technological evolution of ICT and standards*. In: EITO 2000, 2000, pp. 118-217.

Cohen, S., J. Bradford DeLong, J. Zysman, *Tools for thought: What is new and important about the Economy?* BRIE Working Paper 138, Berkeley University, USA, 2000.

CPB, *ICT en de Nederlandse economie – Een historisch en internationaal perspectief*, Werkdocument 125. Centraal Planbureau, Den Haag, 2000.

Ducatel, K., Burgelman, J-C, Howelss, J. Bohlin, E., Ottisch, M., *Information and Communication Technologies and the Information Society Panel Report*. Futures Report Series 03. Sevilla: IPTS, 1999.

Forsebäck, L., *Case Studies of the Information Society and Sustainable Development*. IST-programma, May 2000.

Huiberts, E.J.T. Mobiliteit van Data tegen (w)elke prijs? *Verkennende studie naar het energiegebruik van ICT-infrastructuur voor 2000-2009*. Den Haag, 2001.

ICT and Sustainability, *Making work more environment-friendly and creating a better society*. http://www.sectec.co.uk/wired/bp14.htm, 21 August 2000.

Intel, (2001): http://www.intel.com/research/silicon/mooreslaw.htm

Martens, M.J. et al., *The mobility impact of the electronic highway - A technology forecast and assessment using a scenario approach*. Report 99/NV/074, Delft: TNO-INRO, 1999.

Miles, I., R. Coombs and S. Metcalfe, *Services and innovation*, Background paper for the 6 Countries Programme workshop: Manchester, 22/23 April 1999.

Nederland Digitaal, *Nederland Digitaal – Drie toekomstbeelden voor Nederland in 2030; eindrapportage*. Den Haag: V&W/VROM/EZ, 2000.

Nijhuis, E.W.J.T., G. Scholl, A.F.L. Slob, *Innovation of eco-efficient producer services*; report stage 1: conceptual paper; EU-project 'Creating Eco-efficient Producer services'; IÖW, TNO-STB, Delft, 2000.

Nijhuis, E.W.J.T., *Precisielandbouw en TNO*, TNO Strategie, Technologie en Beleid, Delft, STB 99-13, 1999.

Nijhuis, E.W.J.T., A.B. Smit, S.M. Janssens, *Precisielandbouw biedt akkerbouwer toekomst*, Landbouwmechanisatie, pp. 44-45, January 2000.

Puylaert, H. et al., *Wonen en werken: uit en thuis met ICT – Een verkenning naar de invloed van ICT op (de relatie tussen) wonen en werken*, Rapport P99-007, Delft: TNO-INRO, 1999.

Remote Sensing Nieuwsbrief, *Themanummer Precisielandbouw*, 86, June 1999.

Romm, J. et al., *The Internet Economy and Global Warming*, GEFT/SECS, December 1999. (www.cool-compagnies.org).

Slob, A.F.L., M. J. Bouman, M. de Haan, K. Blok, K. Vringer, *Consumption and the Environment: Analysis of trends*, TNO-STB, University of Utrecht, CBS, Ministry of Housing, Spatial Planning and Environment, the Hague, 1996.

Slob, A.F.L., *Close encounters: Business opportunities in a sustainable society are no science fiction*, paper for the International Business Forum 'Sustainable Consumption and Production' 13 – 15 October 1999, Berlin, TNO Institute of Strategy, Technology and Policy, Delft, 1999.

TLN, *Nieuwe wijn in oude zakken - Pleidooi van Transport en Logistiek Nederland voor meer ruimte voor het vrachtverkeer om de groei van de nieuwe economie op te vangen*. Zoetermeer: Transport en Logistiek Nederland, 2000.

Chapter 9

Economy versus environment? Design alternatives for emissions trading from a lock-in perspective

Edwin Woerdman, Jan-Tjeerd Boom and Andries Nentjes

In this chapter...

...addressing the governance issue, it will be argued that a trend break should be established by using tradable emission rights as the major policy instrument to achieve drastic emission reductions. The design of a tradable system of this type is, however, of crucial importance. The introduction of credit trading is likely to make the current policy framework of standards, covenants, taxes and subsidies more flexible. However, this will result in a 'lock-in' situation which is neither effective nor efficient, because credit trading does not place a cap on total emissions and leaves the entitlement to emit unpriced. Permit trading, on the other hand, is an effective and efficient alternative, because it places a ceiling on total emissions and attaches a price to the entitlement to emit, providing a strong incentive to switch to sustainable energy. This chapter shows that, contrary to common belief, permit trading is not only feasible for industry, but also for consumers, and that administrative costs can be kept low.

Introduction

Reaching a target of reducing CO_2 emissions by 80 per cent in 2050 in The Netherlands requires government policy which is both effective and efficient. This is not just desirable but essential, because ineffective policy instruments which have uncertain effects on CO_2 emissions will not ensure that this ambitious target will be achieved, while inefficient policy instruments are too costly. In broader terms, one could also say that meeting such a drastic emission reduction target calls for reconciliation of the economy (efficiency) and the environment (effectiveness).

Before a policy instrument can be implemented it must be accepted by society (acceptability), for instance by industry and households. Whether an instrument is acceptable depends on its effects on efficiency and effectiveness, its distributional consequences, such as its impact on major interest groups, as well as on the economic, political and cultural context. The characteristics of policy instruments can be analysed as such, but the context in which they will or may operate depends on future developments. This is why earlier in this book two possible futures, or 'visions', were developed to streamline the analysis of long-term national climate-change policy.

Briefly, in market-oriented 'Vision A', society generally prefers a cost-effective climate policy with the assumption, among other things, of a high level of economic growth and technological innovation. In 'Vision B' society primarily prefers an environmentally-effective and equitable climate policy with the assumption, among other things, of a lower level of both economic growth and technological innovation than under 'Vision A'.

In this chapter we will analyse the consequences of different policy instruments under both 'visions'. We will take the plea for emissions trading in Chapter 8 one step further by demonstrating that there are two options, namely credit trading and permit trading, which make long-term national climate policy less costly and more flexible than the current policy mix of standards, taxes, covenants and subsidies. A crucial factor is that the two design alternatives for emissions trading have markedly different consequences in terms of effectiveness, efficiency and acceptability. We will analyse these differences from a 'lock-in' perspective. As far as we know, we are the first to apply this theoretical framework, which was initiated and developed by David (1985) and Arthur (1989), directly to institutional change in the field of environmental policy-making. We will show that, if introduced, credit trading represents a 'lock-in' because it uses existing policy (to put it simply for the moment), whereas the alternative, permit trading, would constitute a 'break-out' because it largely replaces existing policy.

Based on the aforementioned problems and approaches, we have formulated the following central question for this chapter: is it desirable and feasible to introduce permit trading in long-term national climate policy as an alternative for the potential lock-in of credit trading? To answer this question, we have formulated five subsidiary questions:

- how is Dutch environmental policy likely to evolve towards credit trading?
- what would the alternative option of permit trading look like in Dutch environmental policy?
- what are the advantages and disadvantages of permit trading and credit trading?
- what permit trading design could be feasible for industry and households?
- what is the danger of a policy 'lock-in' and how can a 'break-out' be established?

This chapter is organized as follows. The following (second) section describes the historical development of Dutch environmental policy and analyses how it is likely to evolve towards credit trading from a 'lock-in' perspective. The third section presents the general design of permit trading as the alternative option for Dutch environmental policy. Both sections discuss the advantages and disadvantages of credit trading and permit trading. The fourth section presents the details of the alternative option: the so-called downstream permit trading with upstream monitoring approach. The fifth section discusses whether a 'break-out' is possible if Dutch environmental policy finds itself 'locked into' a policy configuration which is unable to actually reach the aforementioned ambitious target at low cost. It also discusses the role of interest groups and gives an overview of recent Dutch policy plans with respect to permit trading and credit trading. Finally, a conclusion is presented.

Possible evolution towards credit trading in Dutch environmental policy

Before sketching the historical development and possible evolution of Dutch environmental policy, we will start by demonstrating how difficult it is to change existing policy arrangements, even if there are better alternatives, on the basis of the 'lock-in' concept. Using this knowledge, we will elaborate upon the advantages and disadvantages of the plausible evolution path towards credit trading. We will also warn against the potential danger of sub-optimal emissions-trading variants being locked into environmental policy.

Theoretical framework: path-dependent development and lock-in

The idea of path-dependent development towards, and lock-in of, sub-optimal situations was introduced in the literature by David (1985), who used these notions to explain technological developments, as described in Chapter 8. Arthur (1989) refined David's concepts to explain (slow) economic change. Both authors argued against the so-called dominant design models. These predict that the optimal design will ultimately win, although in practice a sub-optimal design is often implemented and used during a particular period (such as the inefficient QWERTY arrangement of keys on the typewriter), which can largely be explained on the basis of historical circumstances and learning effects.

Some have already considered the possibility of social lock-in (e.g. Windrum, 1999) or institutional lock-in (e.g. Unruh, 2000), but only to the extent that they reinforce a technological lock-in. In this chapter, we abandon this link with technologies, as suggested by North (1990), and apply the idea of social or institutional lock-in directly to predict, explain and criticize the possible evolution of Dutch climate change policy.[1]

The observation that institutions evolve slowly and that institutional change is difficult has been made in different scientific traditions under different conceptual frameworks, such as the theory of incrementalism (e.g. Lindblom, 1959), the concept of political barriers (e.g. Bachrach and Baratz, 1962), the idea of institutional stability (e.g. North, 1990) and the notion of lock-in (e.g. Arthur, 1994). A sub-optimal policy becomes 'locked in' due to self-reinforcing mechanisms, such as learning effects (e.g. Arthur, 1989), high transition or switching costs (e.g. Katz and Shapiro, 1985) and the vested interests of lobby groups (e.g. Olson, 1982). In this chapter, we use the term lock-in to refer to the dominance of a sub-optimal, in our case inefficient or ineffective, (institutional arrangement of) policy instrument(s) in the system under consideration. We use the efficiency and effectiveness criteria because of the economic and environmental orientation of this chapter.

A dominant or prevailing policy is not necessarily better than its alternatives (cf. Arthur, 1994, p27). Accordingly, we will demonstrate that credit trading, although likely to occur in a traditional, standard-based environmental policy arrangement, is less effective and less efficient than permit trading. It is difficult

---

[1] We kindly thank Bert Steenge from the University of Twente in Enschede, The Netherlands, for making us familiar with North's (1990) idea to apply the technological lock-in concept to institutional (environmental) change.

for a sub-optimal policy configuration to escape from lock-in via evolution and adjustments. Escape is more likely to be facilitated by a break, for instance if an external shock occurs (cf. Arthur, 1994, p129) or if the problem-solving capacity of the dominant policy suddenly declines (cf. Windrum, 1999, p12), while a superior alternative is developed.

Historical development of Dutch environmental policy

We will now consider how environmental regulation developed historically and try to predict how it might evolve into a lock-in situation if policy makers stick to the tried and trusted methods of the past. We will do this by examining the case of The Netherlands.

In The Netherlands, direct regulation was established in 1875 with the introduction of the Nuisance Act ('Hinderwet'). To prevent excessive nuisance for the neighbourhood, industrial plants had to request an environmental licence before they could start their production activities. The new environmental policy of the 20th century, which has grown up since the 1960s, was crafted on this paradigm, which was intended, and indeed apt, to contain only local and simple environmental problems. Experience with direct regulation in the last decades of the 20th century showed that the system worked slowly and involved high administrative costs, while it was sometimes badly co-ordinated and weakly enforced. During those years a lot of adjustments were made to reduce the bottlenecks of direct regulation. For instance, the government has reduced administrative costs, for instance by replacing specific local standards for small sources with national uniform standards, and improving co-ordination by replacing several compartmental environmental laws with a single Environmental Management Act ('Wet Milieubeheer'). The control costs increased as the national environmental targets became more stringent.

The biggest innovation within (and now a typical characteristic of) the established Dutch environmental policy framework has been the introduction of the voluntary agreement, or environmental covenant. The environmental authority and the polluting firm or group of firms discuss how the emission target set by the authority can be implemented in the most flexible and environmentally-friendly way. If an agreement is reached, the terms of the 'contract' are incorporated in the environmental licences of individual firms.

Serious efforts by industrialized countries to contain greenhouse gas (GHG) emissions started in the 1990s after the pledges made in and shortly after Rio de Janeiro (1992) to stabilize CO_2 emissions in 2000 at the 1990 level. The Netherlands aimed to reduce its CO_2 emissions by 3 per cent in 2000 compared with the 1990 level. The national instruments that had been developed by the end of 2000 can be roughly arranged into four broad categories:

(a) energy-efficiency standards, mainly insulation standards for new buildings;
(b) energy-conservation covenants with firms as well as a benchmarking covenant on energy-efficiency with the energy-intensive industry;
(c) a CO_2 emissions tax for small energy users;
(d) subsidies and tax exemptions for energy conservation.

Although this policy mix has probably contributed to tempering the growth of emissions, it is not very transparent and has not been effective in realizing the relatively modest emission stabilization goals for 2000. Quite the contrary: by the year 2000 the greenhouse gas emissions in the growing economy of The Netherlands had risen by about 10 per cent relative to 1990 emissions (COM, 2000a).

Possible evolution of Dutch environmental policy

A major instrument for controlling greenhouse gas emissions by industry and horticulture is the covenant. A feature of Dutch covenants is that they do not set absolute ceilings for emissions, but are clad in the form of a performance standard at the firm level. For CO_2 emissions this means an energy-efficiency standard. The negotiations which precede the covenant offer the participants the possibility to try to reduce total control costs by allocating the highest relative emission reduction to firms with the lowest control costs. This reallocation, and consequently the mitigation of the inflexibility of direct regulation, is constrained, however, since up to now no financial compensation has been provided for firms that take over emission reductions from others.

A logical and incremental next step would be to allow and accept such financial compensation between parties participating in the covenant. It would be an incentive for firms with low (marginal) control costs to accept a higher share of the reduction of emissions. With just a small adjustment this mechanism of financial compensation could evolve into a credit trading scheme. In such a scheme, the current environmental policy targets for the firm are taken as the baseline. Firms that are able to invest in a reduction of their own emissions below the baseline generate credits. These credits can be sold ('traded') to other polluters who will buy them if that is cheaper than reducing their own emissions. Firms may thus use the credits purchased as emission reductions towards meeting their targets. An essential condition for the implementation of such a scheme is that there is a transparent and absolutely certain initial distribution of emissions among the firms participating in credit trading.

To illustrate the plausibility of such a development outside the field of climate policy, the Dutch Minister of the Environment in fact submitted a proposal for NO_x credit trading to parliament in February 2001. The target is to bring NO_x emissions down from 120 kilotonnes in 1995 to 55 kilotonnes in 2010. To reach this goal, the national scheme sets mandatory performance standards of 50 grammes of NO_x emissions per gigajoule of energy input in 2010 for about 200 firms. Credit trading among them is allowed to introduce flexibility. Each year the permissible emissions for a facility are the result of multiplying the performance standard for that year by the energy use of that facility. Although the performance standards may be tightened up when the scheme is evaluated in 2006 (for instance to 40 grammes), the set-up implies that the actual NO_x emission performance of industry as a whole depends on and moves with total energy use.

However, with respect to controlling CO_2 emissions we will argue that an evolution towards credit trading is sub-optimal and leads to a lock-in situation. Its problems stem from the fact that credit trading is based on performance standards or taxes and incorporates not only their advantages but also their disadvantages in a market-based evolution. First, standards, covenants or command-and-control instruments such as energy-efficiency requirements do not place a cap on total emissions, but rather grant emission rights for free to newcomers and allow emission growth to firms if they expand their production. Furthermore, standards are inefficient because emissions are not necessarily reduced where it is cheapest to do so, at the source. Second, although taxes do possess the property of efficiency, their effectiveness is uncertain because policy makers do not know how high taxes have to be to meet the environmental target. Third, if the covenant remains the dominant instrument, periodic negotiations will be required to bring about differentiated standards. The task of establishing environmentally-effective standards in time is likely to be a highly difficult political operation because the government will be permanently confronted with the resistance of interest groups. Fourth, introducing credit trading in the standard-based environmental policy implies that no price is assigned to scarcity. The consequence is that the incentive to restructure production, by substituting sustainable energy for fossil fuels, is too weak.

That final observation can be elaborated. In the current Dutch approach, firms that expand and others that enter an industry are licensed to produce new emissions if they meet the standards. They get the emissions for free (apart from the cost of an environmental licence) – credit trading does not affect this observation. Only in so far as a single firm wants to exceed its licensed emissions does it have to pay for the credits it needs. These expenditures will be a part of the firm's costs and will increase the price of its products. However, for every firm buying credits there will be another firm selling them, hence earning additional revenue and allowing it to lower its product price. So for firms using fossil fuels as a whole, the net impact of the transfer between firms is zero. Basically, from a macro-economic point of view the entitlement to emit remains unpriced under credit trading since scarcity of licences to emit is not reflected in a price for every unit of emission.

Permit trading as an alternative option in Dutch environmental policy

In this section we will show why the alternative of permit trading is more effective and efficient than a lock-in situation of standard-based environmental policy which evolves into a credit trading scheme. We will also indicate how a permit trading scheme should be designed to function properly.

Permit trading versus credit trading in Vision A and Vision B

Permit trading is based on the idea, which would be a trend break, of introducing property (or user) rights in environmental policy (Dales, 1968). These rights are also implicitly allocated in current environmental policy, but permit trading makes them explicit (Tietenberg, 1999).

The permits define the maximum permissible emission levels for each participant. Ideally, these emission levels are derived from the national emission target, so that the individual as well as the national targets are reached. By contrast with a credit trading scheme, the emissions do not increase if the economy grows.

Even when polluters receive their allowances for free ('grandfathering'), expanding firms and market entrants can only acquire their allowances by buying them, thus reducing the opportunities to emit elsewhere in society. Actual emissions and ownership of emission allowances are diligently monitored and compliance is tightly enforced. One can thus expect permit trading to have a better chance of achieving the ambitious goal for emission reduction than credit trading.

Permit trading is not only effective, but also efficient (Montgomery, 1972). Trade in the emission entitlements is free. Every unit of emission will have a price, since each unit can be sold. The gross price of fossil fuels will include a surcharge for the associated emissions and scarcity of environmental space will be reflected in a higher price for fossil fuel-intensive products. This will also stimulate techno-logical innovation, such as cleaner cars or cleaner boilers for households. The process of economic restructuring towards sustainable energy use is thus enhanced in an efficient way. From an efficiency point of view, given the structure of production, the basic difference between credit trading and permit trading is not so much the minimization of emission control costs, since both schemes tend to make marginal control costs equal in the sectors in which they are applied. Rather, the two schemes differ in how they affect the price of fossil fuel-intensive output and the structure of production. Permit trading under an emissions cap forces a more efficient restructuring of production.

Because of its theoretical properties of *both* efficiency and effectiveness, per-mit trading satisfies society's preferences both under market-oriented 'Vision A' and under nature-oriented 'Vision B'. Moreover, these properties of permit trading imply a reconciliation of economy and environment and thus help to solve the dilemma of climate change versus sustainability referred to in Chapter 3. Nevertheless, permit trading is most viable under market-oriented 'Vision A'. Permit trading will be less acceptable (despite its effectiveness) under 'Vision B' because one of the assumed characteristics in that vision (see Table 2.1 in Chapter 2) is a cultural scepticism towards the role of market mechanisms in solving envi-ronmental problems. This means that the political barriers to permit trading, and hence also the costs of establishing such a system, would be lower in 'Vision A' than in 'Vision B'. Some crucial differences between credit trading and permit trading are summarized in Table 9.1.

Taxes are a cost-effective alternative under 'Vision A', but they seem less polit-ically acceptable than permit trading. Emissions trading allows for grandfathering of (part of) the permits, which amounts to giving them away for free. Industry then only has to pay for its reductions, not for its emissions as with taxes. In addition, permit trading makes explicit use of market mechanisms, which are presumably preferred by the society portrayed in 'Vision A'. Furthermore, policy makers do not know how stringent the tax level must be to reach the environmental target. Consequently, ensuring effectiveness in a situation of high economic growth and energy use as assumed in 'Vision A' (see Table 2.2 in Chapter 2) requires trading under an emission ceiling rather than taxes.

At first sight, standards could be seen as an environmentally-effective alterna-tive for emissions trading under 'Vision B', but that society's economy cannot afford the drastic reductions envisioned unless it lowers its costs by using eco-nomic instruments, in particular by means of (grandfathered) permits (which save

Table 9.1 *Crucial (not all) differences between credit trading and permit trading*

	Emissions trading	
Characteristics	Credit trading	Permit trading
Economics	- Cost-savings	- Cost-savings
	- Inefficient: scarcity unpriced	- Efficient: scarcity priced
Environment	- Effectiveness uncertain: no emission ceiling	- Effective: fixed emission ceiling
	- Reduction calculated from baseline emissions	- Reduction calculated from emission ceiling
Politics	- Industry: advantages are cost-savings, no emission ceiling and gratis allocation of emissions	- Industry: advantages are cost-savings and gratis allocation of emissions, but disadvantages are emission ceiling and possible auctioning of emissions
	- Government: advantage is cost-savings, but disadvantages are no emission ceiling and no revenues from gratis allocation	- Government: advantages are cost-savings, emission ceiling and revenues if auctioning, but disadvantage is explicit allocation of property (or user) rights
Visions	- Acceptability likely in regulation-oriented Vision B	- Acceptability likely in market-oriented Vision A
Dilemmas	- Dilemma climate change versus sustainability not solved (scarcity unpriced: switch to sustainable energy is weak)	- Dilemma climate change versus sustainability solved (scarcity priced: incentive to switch to sustainable energy is strong)

permit expenditures). Nevertheless, permit trading is likely to be less politically acceptable than taxes under 'Vision B', because the society's culture is assumed to be more sceptical towards, and concerned about, the use of market mechanisms.

The general design of a permit trading scheme

In order to function effectively and efficiently, a permit trading scheme must be designed properly. The essential design characteristics of a tradeable permit scheme in which the participants can trade under an emissions cap can be summarized in five aspects:

(1) *'what' – permit definition and effectiveness*
The government defines a unit of pollution. A tradeable permit is the right to emit a unit of pollution. For instance, one carbon permit allows the holder to use a quantity of fossil fuel containing one tonne of carbon. To ensure effectiveness, each firm or household in the programme is only allowed to emit units of pollution if it is in possession of a sufficient number of permits. At the end of the year the participant has to turn over to the supervisory authority a number of permits equal to its emissions.

(2) *'when' / 'where' / 'who' – time-path and sector coverage*
The government decides on the economic sectors to be covered by the pro-grammes and on the time-path of maximum tolerated pollution per year for a long future period. The number of permits issued per year is equal to the maximum level of pollution units allowed. The government does not have to buy the permits from the market, as some would think, because the annual permissible level is not only allocated for several years to come, but also declines each year (if a reduction is envisaged).

(3) *'how' – permit allocation*
There are two alternative (or complementary) types of permit allocation: (a) per-mits are sold by the government to the highest bidders, and/or (b) permits are dis-tributed for free among the established polluters (grandfathering). To overcome opposition from fuel-intensive sectors, experience suggests that grandfathering based on historical use, implicitly brought about by granting rights, is politically the most expedient way of initial distribution (e.g. the US SO_2 allowance scheme). In this section we therefore concentrate on a scheme where permits are grandfa-thered. The government only decides how to allocate the permits to start the pro-gramme. How emissions are allocated among sources during the programme is left to the market participants themselves.

(4) *'market' – trade and efficiency*
To ensure efficiency, the permits can be traded. A source of pollution (e.g. firms and households) that wants to emit more than allowed by the permits it owns can buy additional permits; a source of pollution that manages to emit less than the permits in its possession is allowed to sell the surplus. Banking is allowed and the carbon permit can be used at any time.

(5) *'government' – monitoring and enforcement*
Without interfering with the permit transactions themselves, the government monitors and enforces the permissible emission levels. The (national) scheme can only function well if the following conditions are satisfied:
(a) binding national targets and timetables;
(b) reliable national registration of individual emission sources;
(c) adequate monitoring of emissions (either directly or indirectly);
(d) precise registration of permit ownership and permit transfers;
(e) compliance checks by comparing permit ownership and actual emissions;
(f) effective enforcement by applying sanctions in case of non-compliance;
(g) application of existing anti-trust legislation to the permit market.

A design for downstream permit trading with upstream monitoring

Permit trading should focus on restricting fuel use. The first reason is that, for the time being, fuel saving and fuel substitution are the principal, and almost only, economically feasible options for reducing emissions of CO_2. The second reason relates to administrative efficiency and enforcement. The discussion on the design of a domestic permit-trading scheme usually focuses on the choice between a downstream or upstream trading system. In a downstream trading system both large and small energy end-users, including industry as well as households and motorists, receive tradeable permits. The common view seems to be that such a large number of traders would be administratively unfeasible and costly (cf. Hargrave, 1999). An upstream trading system is then thought to be more acceptable, in which the permits are allocated to producers and importers of fossil fuels who pass on their permit costs in a mark-up on the fuel price for both large and small emitters.

However, we challenge this view, arguing that a downstream system which directly incorporates firms as well as households and motorists can indeed be administratively feasible and cheap, thereby enhancing its potential acceptability. This can be done, among other things, by concentrating the monitoring activities as far as possible on fossil fuel producers and importers (upstream) and by using 'chip card' technology for households and motorists (downstream). The energy demand from transport, for instance, is projected to become higher than from industry, not only in 'Vision A', but in particular in 'Vision B' (see Table 2.2 in Chapter 2). Our message is that a domestic permit trading system is not only feasible for industry, but also for consumers. The outline of such a downstream trading and upstream monitoring approach is sketched below.

Using bookkeeping of fuel sales to incorporate industry

Carbon permits are grandfathered to firms based on the carbon use in a reference year or the average over a number of reference years. Gas and electricity use can be measured on the basis of the payments to, and annual receipts from, the distribution companies. Commercial use of fossil fuel for driving a car owned by a firm has to be demonstrated via the bookkeeping of the firm. Companies that remain below a certain threshold of fuel use can be allocated permits on the basis of a generic criterion, such as the number of employees (in FTEs). In this scheme of grandfathering, some of the permits could be withheld and auctioned once or twice a year to accommodate the fear that newcomers and expanding firms would be unable to buy a sufficient number of permits.

Purchasers of fuels have to pay with permits. Distributors are forbidden from selling fuels except in return for a permit. For every tonne of fossil fuel a firm or household purchases from distributors, it has to hand over an equivalent number of carbon permits. For their part, distributors can only obtain fuels from their suppliers in exchange for carbon permits. In this way all permits will end up in the hands of producers and importers of fuel, including the permits purchased by distributors to cover their fuel supply to consumers and other small users. Producers and importers of fuel are then obliged to turn over to the environmental authorities carbon permits for the carbon contained in the fossil fuels they have sold on the market.

Monitoring of emissions (fuel sales) and checking whether they correspond with permits can be concentrated upstream on producers and importers of fuel. Their number is limited: in The Netherlands about 40 to 50. The bookkeeping of fuel producers and importers is checked at the end of the emission year to determine how many permits are actually present and how many there should be by calculating the number of required permits on the basis of the administration of fuel sales. Perhaps the administrative monitoring can be (partially) privatized by requiring an audit certificate from an accountant. If it is found to have a shortage of permits, the fuel producer or importer is given one month to obtain (in other words, buy) the necessary permits. If it is unable or unwilling to do so, the company receives a fine which is many times higher than the highest expected market price of the permits, and is still obliged to hand over the missing permits to the authorities.

The system is to a large extent self-regulating. In this design, fuel producers and importers (as well as distributors) have an interest in receiving the correct number of permits for their fuel sales. The supplier does not want to deliver fuel without receiving permits from the buyer. It is not necessary for the national agency to monitor the millions of fuel users. This keeps the cost of monitoring and enforcement low. The monitoring scheme fits in with existing institutions in most Western countries for levying excise duties on fossil fuels. In The Netherlands, traders and suppliers of mineral oils are presently obliged to have a licence and to report each month the quantity they have supplied to the market, while they have to turn over the excise duties to the authorities. This administrative system of self-reporting is supplemented by occasional physical checks (Koutstaal, 1997).

Using chip cards and electricity bills to incorporate consumers

The implementation, including monitoring and enforcement, is assigned to a national permit agency. All participants are registered with this agency. The implementation is largely computerized using chip-card technology. Permits are not only grandfathered to firms, but also to households. The direct and active involvement of households in climate policy by giving them a tradeable environmental asset with market value also helps to solve the dilemmas of 'decentralized versus centralized decision-making' and 'demystification versus increased environmental awareness' referred to in Chapter 3.

At the beginning of the year the end-users receive the permits ('user rights') for the coming year for both stationary and mobile sources of emissions on their permit account. The agency also sends a chip card. (Instead of using a separate chip card it might be possible to integrate it in existing chip cards from banks.) Households can upgrade the chip card at the expense of their permit account.[2] When purchasing fuels, the end-user has to transfer a number of permits (corresponding to the carbon content of the acquired fuel) to the permit account held by the distributor at the national permit agency. For the mobile sources the transfer occurs by using the chip card which households can fill by debiting their permit account. An alternative is a permit PIN card which enables individuals to transfer permits directly from their own account to the fuel supplier. Individuals can only

---

[2] Some have suggested allocating permits to the distributors, who will then pass on the permit costs in a mark-up on the fuel price, thereby avoiding allocation to households. However, grandfathering is then not likely to be politically acceptable, because it would create a profit for the distributors (companies like Shell etc.), while the consumers pay for the emission reductions. Households are better off if they (instead of the distributors) get permits for free, not only because they receive a wealth transfer, but also because it enables consumers to make a profit by selling permits if they succeed in using less energy and fuel.

replenish the card or withdraw from the account if the permit balance is positive. For stationary sources the transfer of permits is linked to the annual gas and electricity bill from the distribution company. If a household does not have sufficient permits the distributor has the right to buy the required permits, and to recover the costs from the client.

When a motorist goes to a petrol station, he or she not only pays for the fuel but also transfers a number of permits corresponding to the carbon content of the acquired fuel into the permit account which the distributor holds at the national permit agency. Machines are installed at filling stations and at other strategic locations where one can electronically replenish (buy) or withdraw from (sell) the permit chip card at the current market price. The automated machines are operated by companies that trade professionally in carbon permits. The current market price arises from the transactions by and between the permit trading companies. At the end of the year, the national agency establishes the balance of the permit account for every user. This is equal to the grandfathered permits (via chip card or account) plus the purchased permits minus the permits sold and the permits used and transferred. This balance, which can only be positive, is added to the permit account for the next year. The surplus permits can be sold by the account holder or they can be kept as an investment.

While there is price certainty under a tax (in the form of the tax rate), nobody knows how high (or low) the price level will be in a market for tradeable emissions in The Netherlands. There are tentative indications that the price in this domestic market could lie around 20 euros per tonne of CO_2, but estimates roughly range from 10 to 50 euros per tonne of CO_2 depending on which sectors participate and how trading is organized (Boots et al., 2001). European or international trading could lower these prices as cheaper options would then become available. It is possible to introduce some price certainty into the market by creating a maximum permit price. Although this will increase political acceptability, it also represents market interference: fixing a maximum price is undesirable from an economic and environmental point of view, because it means that the permit price may not fully reflect the environmental scarcity.

Assessing the level of administrative costs

The introduction of the permit chip card requires investments in automated machines and a telecommunication network. The investment costs are comparable to the costs of installing a PIN card or chip card system for a bank with millions of account holders. These costs could possibly be shared between the permit agency and the banks if the permit chip card were to be combined with existing chip cards from banks. The large-scale character and intensive use of the machines will result in low costs per transaction. Besides the aforementioned costs of the chip card technology (depreciation, interest and exploitation), there are also the time costs of both motorists and employees at petrol stations, associated with the additional action involved in processing the permits.

The implementation costs consist of the registration of the participants as well as the annual allocation of permits and mailing of chip cards. We roughly estimate these to be several million euros, which implies a few euros per chip card. The monitoring focuses on the limited number of importers and producers of car fuels who already have a detailed administration of their fuel sales for commercial and fiscal reasons. The monitoring costs will therefore be limited to no more than several million euros. To put these figures into perspective, we roughly estimate that at a carbon price of between 25 and 75 euros per tonne, the corresponding value of carbon emissions (35 Mton in 2010) would be about 1,000 to 2,500 million euros. Furthermore, the transaction costs are expected to be low given the large number of participants. Using the brokerage costs of the SO_2 allowance programme in the US as an indicator, we estimate the transaction costs to be no more than 4 per cent of the transaction value.

The large number of traders makes market power unlikely. The political process will certainly induce start-up costs, but these are unavoidable and necessary to reap the environmental and economic benefits of a workable permit trading scheme. Versteege and Vos (1995) estimate that the period needed to prepare a tradeable permit scheme for SO_2 and NO_x emissions for the energy-intensive sectors is seven years. This could also be indicative for a similar scheme including CO_2 emissions. If 10 man-years are devoted annually to the preparation of the scheme, the total preparation costs would be about three million euros.

To summarize, we estimate the administrative costs to be several million euros per year. Whether this is cheap or expensive can only be judged when we know the administrative costs of alternative options to reduce the emissions of mobile and stationary sources. Nevertheless, it is our impression that the administrative costs of a scheme with downstream trading and upstream monitoring can be kept to a low level.

'Lock-in' or 'Break-out'?

We have argued that the possible evolution towards CO_2 credit trading based on performance standards and covenants for industry together with CO_2 taxation and subsidies for small firms, the service sector and households, is a 'lock-in'. We do not deny that credit trading will improve the cost-effectiveness of the current national policy mix of standards and taxes. Nor do we deny that tightening up the performance standards for large emitters, greening the tax system and increasing the tax rates and subsidies for small emitters will make a positive contribution to efforts to meet the emission targets. Moreover, we do not deny – and have in fact emphasized – that credit trading is politically easier to develop than permit trading, since the former builds on the existing policy framework. However, we do stress that:

- the current policy mix has been unable to reach relatively modest emission-reduction goals in the past;
- adjusting energy-efficiency standards, tax rates and subsidy levels in the future involves costly government control and has uncertain effects on CO_2 emissions;
- locking credit trading into this policy framework leaves the environmental effectiveness uncertain (and even unlikely) and leaves environmental scarcity unpriced;
- there is an alternative, permit trading, which reduces the uncertainty of reaching the emission reduction goals (effectiveness), lowers overall compliance costs and gives every unit of emission a price (efficiency).

Is a 'break-out' really an impossibility, as some would think? A domestic transition to an economy-wide permit market with a cap, where all participants can trade with each other, would require two main changes. Firstly, the efficiency norms in covenants and direct regulation would have to be translated into fixed CO_2 emission ceilings for the coming years. Secondly, CO_2 taxation and subsidies would have to be terminated and grandfathered permits introduced instead (as described above), or alternatively, permits would have to be auctioned to distributors as proposed by Koutstaal (1997) and Nentjes et al. (1995). From a technical or economic point of view these changes do not seem to be either complex or costly. The major bottlenecks to introducing domestic permit trading are rather of a political nature, as will be explained below (e.g. Woerdman, 2000a; Woerdman, 2000b).

Some of the main barriers to implementation of an (inter)national scheme are: (1) the allocation of permits could lead to unfair changes of the level playing field as perceived by different emitters, (2) the market for tradeable permits has to be set up, which entails some potential novelties for administrative bodies such as permit tracking, trading rules and liability procedures, (3) there may be cultural resistance to the seeming contradiction of introducing the market via the explicit distribution of property rights in order to counteract the externalities of the market itself, and (4) there may be economic resistance from (energy-intensive) industry to absolute limits on the growth of their greenhouse gas emissions. We will elaborate upon the resistance of industry and other interest groups to permit trading a little later from a lock-in perspective.

Introducing a permit trading market on a European or international scale requires that domestic schemes are established first. If countries satisfy certain minimum requirements, so-called eligibility criteria (for instance with respect to permit definition, monitoring and enforcement), these national schemes can be connected. An illustration of the dilemma 'government oversight versus Europeanization' referred to in Chapter 3 is that this would not only increase the economic potential for cost savings, but would also raise the additional legal and political obstacle of whether or not grandfathering should be seen as state aid under EU law or as an 'actionable' subsidy under World Trade Organization law (Woerdman, 2000b).

Reinforcement of the lock-in situation by interest groups

The lobbying of (powerful) interest groups may reinforce the 'lock-in' of credit trading as a sub-optimal policy instrument. Boom and Svendsen (2000a) show that industry should theoretically prefer performance standards at the national level. Performance standards (such as energy-efficiency requirements) have the advantage for energy-intensive industry that they set a level of emissions per unit of production. Hence, production can be expanded without running into very high abatement costs. Furthermore, with performance standards, a firm only has to pay for the emission reduction and not, as with taxes or auctioned permits, for the remaining emissions. Although grandfathered permits eliminate this particular disadvantage for them, industry rejects them because it would still set an absolute ceiling on their emission growth.[3]

---

[3] Interestingly, the shareholders of these firms prefer grandfathered permits (Dijkstra, 1999). The grandfathered permits have a market price and opportunity costs (when they are used to cover the emissions) which have to be recovered by raising the product price, thereby increasing the firm's profit. This particular positive effect on the firm's profit is absent under performance standards.

The theoretical conjecture that industry does not prefer permit trading is supported by empirical evidence. In a survey of Dutch interest groups, Dijkstra (1999) found that industry indeed prefers measures such as performance standards. Although the survey did not consider the option of voluntary agreements, industry did express a preference for covenants in the comments they gave on the survey. In another survey covering a large part of the industrialized countries, Boom and Svendsen (2000b) found that industry generally prefers voluntary agreements which set performance standards at the national level.

Although these authors did not investigate industry preferences with respect to credit trading at the national level, it is assumed that a preference for instruments other than permit trading implies a potential preference for credit trading, because the latter is compatible with such instruments. Indeed, on an empirical level, some (mainly European) industries have expressed a preference for credit trading with relative targets, whereas some (mainly American) industries have explicitly indicated that they prefer permit trading (Boom and Svendsen, 2000b). However, part of the explanation could be that the latter associate carbon credit trading with the early credit-trading schemes in the US which were subject to many restrictions (cf. Klaassen and Nentjes, 1997).

Interestingly, there are not many environmental organizations around the world that contribute to a 'break-out' either (Boom and Svendsen, 2000a; 2000b). Several green NGOs fear permit trading because it is a market instrument which would, they believe, allow the rich countries and companies to 'buy their way out' instead of reducing the emissions themselves (Woerdman, 2000a). Nevertheless, in a few countries, for instance the US and The Netherlands, there are some NGOs which are positive about permit trading because it would generate the highest level of environmental certainty and effectiveness (Dijkstra, 1999).

The environmental bureaucracy has also played its part in the lock-in. When trying to develop acceptable policy measures, democratic governments – in particular civil servants in the consultative culture in The Netherlands referred to as the 'polder model' – are in practice inclined to take particular account of the preferences of interest groups, most of whom have been, as demonstrated above, against permit trading.[4] Moreover, to some extent, theory predicts that the environmental bureaucracy will try to prevent either a loss of policy control to the market or a decrease in the workload for the bureaucracy (Niskanen, 1994), which means they are likely to prefer performance standards over permit trading at the national level.

Permit trading and credit trading in Dutch policy plans

Policy makers in The Netherlands as well as in other European countries have recently become more optimistic about emissions trading. The European Commission openly acknowledged in a Green Paper that the Kyoto Protocol of

---

[4] The Dutch word 'polder' is characteristic of The Netherlands, because it refers to the low-lying lands behind the dikes. The 'poldermodel' (which has been applauded by Bill Clinton in the US and by Tony Blair in the UK as the 'Third Way') refers to a consultative culture which is characteristic of The Netherlands. Although this culture has a long tradition, it is usually associated with the rather successful socio-economic policy in The Netherlands which originated in the 1980s. From then on, trade unions and employers' organizations increased their co-operation by voluntarily agreeing to moderate wage claims and accepting more privatization in order to strengthen firms' competitive positions. The trade unions hoped that this would generate a positive effect on long-term job creation and wage development for employees.

1997 has put emissions trading on the EU agenda (COM, 2000b: 7). The Netherlands has conducted several serious studies into (inter)national permit trading, in particular since Kyoto. However, there are also proposals for credit trading. It thus remains to be seen whether the studies on permit trading turn out to be the first steps towards a 'break-out'.

In The Netherlands, the ministry of the environment is called the 'Ministerie van Volkshuisvesting, Ruimtelijke Ordening en Milieubeheer' (Ministry for Housing, Spatial Planning and the Environment), abbreviated as VROM, and the ministry of economic affairs is called 'Ministerie van Economische Zaken', abbreviated as EZ. In the late 1990s, an advisory council to the ministry of the environment (VROM Raad, 1998) recommended the introduction of permit trading for a part of the 'sheltered' sector (those that either do not compete on the international market or operate with relatively low energy costs, such as households) and to rely on traditional policy for the 'exposed' sectors (those that are exposed to international competition and energy-intensive production, such as industry) as long as international permit trading is not possible. The Dutch ministry of the environment itself (VROM, 1999b) then went a little further by concluding that a tradeable permit scheme should be implemented to reduce CO_2 emissions, in the short term only for the sheltered sector (VROM, 1999a). The Dutch ministry of economic affairs (EZ, 1999) agreed with the view of the environment ministry that permit trading should in the long run also be developed for all sectors, but added that credit trading should initially be allowed for the exposed sector. Reflecting these negotiating outcomes and the progression of ideas, the Dutch environment ministry installed the Vogtländer Commission (CO_2 Trading Committee), which includes officials, firms and NGOs. In a preliminary report at the beginning of 2001, it suggested creating permit trading for the sheltered sector and credit trading based on performance standards for the exposed sector.[5]

Interestingly, similar to our argument in this chapter, both the environment and the economic affairs ministries acknowledged that permit trading should, in the long term, largely replace traditional policy arrangements (VROM, 1999a, p102; EZ, 1999, p32).[6] In light of the dilemma 'central steering versus co-production' referred to in Chapter 3, it could indeed be argued that a policy mix, as proposed in Chapter 8, is undesirable once property (or user) rights and a market to trade them are created. Any other policy instrument aimed at achieving a CO_2 emission reduction target of 80 per cent in 2050 would disrupt this market and is bound to make it less efficient or effective. Of course, there would be a place for a fuel tax to generate revenue for the government and to realize an equitable and efficient distribution of the tax burden between labour, capital and natural resources. However, the tax should not be set with the aim of regulating emissions, since this is done by the permit price. Subsidies for investment to control emissions would certainly be undesirable: they would distort the permit price sig-

---

[5] The Vogtländer Commission also proposed an emissions exchange between these two sectors which basically lifts the emission ceiling of the sheltered sector, but it should be noted that its final report (which is expected in December 2001) was not yet finished when this chapter was written.

[6] 'Veel van de bestaande instrumenten zullen overbodig worden of slechts een ondersteunende rol krijgen' ['Many of the existing instruments will be superfluous or have only a subordinate role'] (VROM, 1999a: 102). 'Toepassing van verhandelbare emissies is alleen zinvol wanneer dit het bestaande beleid zo veel mogelijk vervangt' [The application of tradeable emissions is only useful if it so far as possible replaces existing policy'] (EZ, 1999: 32).

nal. Emission standards might be used as a method to set baselines for credit trading and voluntary agreements to establish emission ceilings for target groups of industries.

In the meantime, at the European level the EU produced a Green Paper which foresees the start of EU-wide experimental permit trading by 2005 (COM, 2000b). By contrast with the Dutch approach, the EU suggests introducing permit trading first among large (rather than small) emitters, the exposed (rather than the sheltered) sectors if you will, such as electricity producers. It also proposes maintaining standards for the household and transportation sectors, at least initially, and studying the possibility of credit trading further.

Before the installation of the Vogtländer Commission in The Netherlands, the authoritative Social and Economic Council (SER), whose (legal) role is to advise the Dutch government, argued in favour of permit trading and was the first body in The Netherlands to advise *against* credit trading because of its disadvantages (SER, 2000), which were discussed at length in the previous sections. It goes without saying that this chapter supports the views held by the SER. Nevertheless, we have taken the argument one step further by warning policy makers against implementing a sub-optimal credit trading scheme because of the potential danger of lock-in.

Politically, a 'break-out' to permit trading will not be easy. However, there are at least two reasons from a political-economy perspective why it is easier to go from standards directly to permit trading than to go from standards via credit trading to permit trading. First, although credit trading is less effective and efficient, it does capture much of the potential cost saving from trade, making the economic advantages of a change to permit trading appear smaller and less visible. Second, going from standards to credit trading locks in the vested interests of the lobby groups discussed above, who will then be less willing to accept permit trading, precisely because they already enjoy the fruits of flexibility while avoiding an emissions cap. However, credit trading should not be locked into the traditional policy framework because permit trading is more environmentally effective and efficient. Therefore, society and government are better off if a trend break in climate policy is established towards permit trading.

Conclusion

This chapter demonstrates that it is possible to reduce CO_2 emissions by 80 per cent in 2050 in The Netherlands provided that a trend break is made away from the kind of policy instruments now used. The history of environmental and climate policy in The Netherlands has led to a policy framework which is mainly built on standards, covenants, taxes and subsidies. If this framework is to be made more flexible in order to afford the envisaged 80 per cent emission reduction, it is likely to show an evolution towards credit trading, because that system uses these traditional instruments as the baseline from which to calculate and trade emission reductions.

However, the current policy framework has been unable to reach the relatively modest emission reduction goals in the past and it will have uncertain effects on

CO_2 emissions in the future. Firstly, a scheme of performance standards supplemented with credit trading does not place a cap on total emissions, but rather allows emission growth to firms if they expand their production. Secondly, credit trading leaves the entitlement to emit unpriced, so that there is too little incentive to restructure production and substitute fossil fuels with sustainable energy. We refer to this plausible path-dependent incorporation of credit trading into the current policy framework as a 'lock-in', because it results in a sub-optimal situation which is neither effective nor efficient.

A trend break in terms of policy instruments is desirable, or even essential, because there is a more effective and efficient design alternative for Dutch environmental policy. This alternative is permit trading. Firstly, permit trading explicitly allocates property (or user) rights and places a ceiling on total emissions. This means that, contrary to credit trading, effectiveness is also achieved if firms increase their production. Furthermore, permit trading not only leads to cost savings, but also attaches a price to the entitlement to emit, so that efficiency is ensured and the incentive to switch to sustainable energy is strong. In our view, permit trading would thus reconcile economy and environment and solve the dilemma of climate change versus sustainability.

We emphasize that a domestic permit trading system is not only feasible for industry, but also for small emitters like households and car drivers. We have proposed a permit scheme with downstream trading and upstream monitoring. We show that, contrary to common belief, administrative costs can be kept low by distributing the permits and concentrating monitoring on fossil fuel producers and importers (upstream), and by using chip card technology for households and motorists (downstream). For households, the permit transfer would be linked to the annual gas and electricity bill from the distribution company. For motorists, the permit transfer would occur by using a chip card which they can replenish from their permit account. The direct involvement of consumers would also help to increase awareness of climate change.

This particular design, which combines downstream and upstream elements, is technically feasible and economically desirable. Whether it is also politically acceptable, leading to a trend break (or 'break-out'), depends on several factors, particularly the willingness of industry to accept emission limits and the willingness of households to accept market instruments for environmental policy. Market-oriented 'Vision A' seems to offer better prospects for permit trading than 'Vision B', which is more closely associated with standards and taxes.

A permit scheme with downstream trading and upstream monitoring would not only incorporate elements of decentralization and government oversight, but also of Europeanization. The Kyoto Protocol already put emissions trading on the European agenda. It is not unthinkable that the current development of permit trading schemes in Denmark and the UK (as well as in the US) will provide experience with, and confidence in, the instrument, stimulating a similar development in other EU member states. The Netherlands would be particularly suited to make an early switch to permit trading, because of its aspirations for environmental leadership, its history as a trading nation and the quality of its monitoring and enforcement structures.

References

Arthur, W.B., *Increasing Returns and Path Dependence in the Economy*, Michigan, The University of Michigan Press, 1994.

Arthur, W.B., Competing Technologies, Increasing Returns, and Lock-In by Historical Small Events, *Economic Journal* 99, 1989, pp. 116-131.

Bachrach, P. and M.S. Baratz, Two Faces of Power, *American Political Science Review* 56 (4), 1962, pp. 947-952.

Boom, J.T. and G.T. Svendsen, The Political Economy of International Emission Trading Scheme Choice: A Theoretical Analysis, *Journal of Institutional and Theoretical Economics* 156 (4), 2000a, pp. 548-566.

Boom, J.T. and G.T. Svendsen, *The Political Economy of International Emissions Trading Scheme Choice: Empirical Evidence*, Mimeo, Department of Economics, Odense, University of Southern Denmark, 2000b.

Boots, M., M. de Feber, M. Koster and A. Wals, *Verhandelbare systemen in het Nederlandse klimaatbeleid*, Report Nr. ECN-C-01-044, Petten, Energie Centrum Nederland, 2001.

COM, *EU Policies and Measures to Reduce Greenhouse Gas Emissions: Towards a European Climate Change Programme*, Communication from the Commission to the Council and the European Parliament, document COM(2000) 88, 8 March 2000, 2000a.

COM, *Green Paper on Greenhouse Gas Emissions Trading Within the European Union*, Green Paper presented by the Commission, 8 March, 2000b.

Dales, J.H., *Pollution, Property and Prices: An Essay in Policy-Making and Economics*, Toronto, Toronto University Press, 1968.

David, P., Clio and the Economics of QWERTY, *American Economic Review* 75, Papers and Proceedings, 1985, pp. 332-336.

Dijkstra, B.R., *The Political Economy of Environmental Policy: A Public Choice Approach to Market Instruments*, Cheltenham, Edward Elgar, 1999.

EZ, *Milieu in de Markt: Emissiehandel als beleidsinstrument*, Den Haag, EZ Ministry for Economic Affairs, 1999 (in Dutch).

Hargrave, T., *Identifying the Proper Incidence of Regulation in a European Union Greenhouse Gas Emissions Allowance Trading System*, Draft, June 1999, Washington, Center for Clean Air Policy (CCAP), 1999.

Katz, M. and C. Shapiro, Network Externalities, Competition and Compatibility, *American Economic Review* 75, 1985, pp. 424-440.

Klaassen, G., and A. Nentjes, Sulfur Trading Under the 1990 CAAA in the US: An Assessment of the First Experiences, *Journal of Institutional and Theoretical Economics* 153 (2), 1997, pp. 384-410.

Koutstaal, P.R., *Economic Policy and Climate Change: Tradeable Permits for Reducing Carbon Emissions*, Cheltenham, Edward Elgar, 1997.

Lindblom, C.E., The Science of 'Muddling Through', Indianapolis, Bobbs-Merrill, 1959.

Montgomery, W.D., Markets in Licences and Efficient Pollution Control Programs, *Journal of Economic Theory* 5 (3), 1972, pp. 395-418.

Nentjes, A., P.R. Koutstaal and G. Klaassen, *Tradeable Carbon Permits: Feasibility, Experiences, Bottlenecks*, NRP Report No. 410100104, Dutch National Research Programme (NRP) on Global Air Pollution and Climate Change, Bilthoven, NRP, 1995.

Niskanen, W.A., *Bureaucracy and Public Economics*, Second Edition, John Locke Series, Aldershot, Edward Elgar, 1994.

North, D.C., *Institutions, Institutional Change and Economic Performance*, Cambridge, Cambridge University Press, 1990.

Olson, M., *The Logic of Collective Action: Public Goods and the Theory of Groups*, Second Edition, Cambridge, Harvard University Press, 1982.

Tietenberg, T., Lessons From Using Transferable Permits to Control Air Pollution in the United States, in: J.C.J.M. van den Bergh (ed.), *Handbook of Environmental and Resource Economics*, Cheltenham, Edward Elgar, 1999, pp. 275-292.

Unruh, G.C., Understanding Carbon Lock-In, *Energy Policy* 28, 2000, pp. 817-830.

Versteege, H.M.J. and J.B. Vos, *Verhandelbare emissierechten in het Nederlandse Verzuringsbeleid*, DHV Rapport Nr. 116, Den Haag, VROM Ministry for Housing, Regional Development and the Environment, 1995 (in Dutch).

VROM, *Uitvoeringsnota Klimaatbeleid – Deel I: Binnenlandse Maatregelen*, Den Haag, VROM Ministry for Housing, Regional Development and the Environment, 1999a (in Dutch).

VROM, *Verhandelbare emissies als instrument in het milieubeleid*, Brief 26578(1) van J.P. Pronk aan de voorzitter van de Tweede Kamer der Staten-Generaal, Den Haag, VROM Ministry for Housing, Regional Development and the Environment, 1999b (in Dutch).

VROM-Raad, *Transitie naar een Koolstofarme Energiehuishouding: Advies ten Behoeve van de Uitvoeringsnota Klimaatbeleid*, Advies 010, Den Haag, VROM-Raad, 1998 (in Dutch).

Windrum, P., *Unlocking a Lock-In: Towards a Model of Technological Succession*, Maastricht, MERIT / University of Maastricht, 1999.

Woerdman, E., Organizing Emissions Trading: The Barrier of Domestic Permit Allocation, *Energy Policy* 28 (9), 2000a, pp. 613-623.

Woerdman, E., Competitive Distortions in an International Emissions Trading Market, *Mitigation and Adaptation Strategies for Global Change* 5 (4), 2000b, pp. 337-360.

Chapter 10

Legal aspects of a changing energy system in The Netherlands in 2050

Henk Addink

In this chapter...

... the legal dimensions, including fundamental norms, of the required societal changes towards a climate-neutral society are examined systematically and presented from a normative-instrumental perspective. These legal aspects have essentially been ignored in much of the thinking about long-term climate policies. This has also been the case in the recent Netherlands' Climate Change Policy Implementation Plans. The focus in this chapter is on relevant elements in the field of law: institutional responsibility, instruments and norms, enforcement, public participation and legal protection. Illustrations of each of these elements are also given. This chapter sets out the legal elements and conditions for the development of a successful Netherlands long term policy on climate change. These conditions should be integrated in the daily climate change policy of The Netherlands.

Introduction

In this book, radical changes in the energy system in The Netherlands are explored over the long term. Radical changes are necessary because of the need for extreme reductions of greenhous gas emissions. In the interest of protecting the environment, a considerable reduction of CO_2 emissions is especially necessary. This means that the energy system of The Netherlands has to be transformed to use less CO_2-emitting energy.

It is of great importance for policymakers to stipulate that the energy system per se does not determine the reductions. This is also relevant from a legal perspective because in this kind of situation, legal precedent – both national and international – shows two separate approaches: one from the perspective of energy law and one from the environmental law side. However, an intersectoral approach is essential and makes integration in the policy field possible. This would also be the answer for the dilemma of 'climate change versus sustainability' discussed in Chapter 3.

Unlike the situation in the 1960s and '70s, when there was an oil crisis and discussions about the limits of growth in relation to the availability of raw materials, nowadays there are no real problems with the availability of energy resources. Instead, from the perspective of the energy market, there are external factors, such as climate change, which have led to the conclusion that a reduction in CO_2 emissions is necessary. This has created a situation with serious consequences for the energy market. The basis for this change is to protect the environment, which is the focus of environmental law. This means that there must be a break from current trends, along with the development of market-related instruments to measure changes in the supply and demand of energy resources. However, the answer does not only – not even primarily – lie in the creation of covenants and trading permits, which would not be enough to address the fundamental norms such as rule of law, principles of good governance, democracy and human rights. Relations with the economic market can also be made by using public law instruments.

The legal approach for the visions of the future

As discussed in Chapter 2, the problem can be approached using visions A and B, which give two realities for the year 2050. The basis of this approach is essentially socio-economic, focusing on the tension between economic and environmental aspects of the situation. This has been covered in Chapter 9, particularly in the discussion of permits and credit trading, in which there was no attention paid to the fundamental legal questions. However, if policymakers want to develop a climate policy that can be effectively and successfully implemented, there are legal questions that must be addressed. It is these 'missing' legal questions that are discussed in this chapter by answering the following question: what intersectoral legal conditions (in relation to the two theoretical visions, A and B) must be considered if the substantial break with current trends is to be realized?

This legal approach essentially consists of three parts:

1. Who? (i.e., the persons/institutions involved)
2. How? (i.e., the instrumental framework)
3. What? (i.e., the relevant norms).

The legal-normative aspects to be worked out in this context need to be distinguished from the ethical-normative and the moral-normative aspects. The essential difference between the latter two and the legal-normative aspects is that the legal-normative aspects can be enforced legally, which is a pre-condition for a successful climate policy.

Another relevant part of the legal approach is the goal and the function of the legal norms, in other words, the purpose of a norm and the choice of the (legal) instrument that is related to it. It is clear that to establish this type of norm, certain goals have to be reached. This is the instrumental side of the legal approach, through which a direct link can be made with public administration and policy science. When I refer here to 'administration', 'administrative functions' and 'administrative control', I mean 'public administration', 'public administration functions' and 'public administrative control'.

Norms and instruments in the legal context are always related to persons and institutions. These three elements – persons/institutions, instruments and norms – together are the cornerstones for the 'instrumental-normative legal approach', which is the basis of the intersectoral legal discussion.

The legal conditions necessary to break the trend in climate change by the year 2050 are discussed below. But first, we must look at the problem in The Netherlands now and the importance of policy. It is in this context that approaches for the long term must be developed. Then we discuss the instrumental-normative legal approach in relation to the problem of reducing greenhouse gases. This begins with a few legal themes or starting points that are essential to clarifying the legal dimensions of the problem. These provide 'checkpoints' for developing a legal perspective of the visions A and B. Next the two visions are discussed in relation to these 'checkpoints'. Each vision is described briefly, the elements of the legal approach for each vision are discussed, and finally, the elements are integrated. We then compare the two visions, based on the legal elements, with possible outcomes projected, after which we link the opinion of the Dutch Minister of Environment on climate change policies on the prospects for the long term. This paper concludes with conclusions and the recommendations.

Long term prospects in Dutch climate change policy

The importance of the problem is recognized not only in theoretical but also in practical policy fields. In Chapter 4 of the 1999 Netherlands' Climate Policy Implementation Plan, it is stated that, in order to be able to breach the trend and actually reduce CO_2 emissions over time, the foundations for a long-term policy must be laid down. The Minister endorses the advice of the Environmental Advice Council, which considers technological and instrumental innovation a prerequisite for the transition to an economy based on low-carbon energy and believes that such a transition is essential if The Netherlands is to continue to participate in international climate policy.

From the Minister's point of view, greenhouse gas emissions will ultimately have to be drastically reduced to prevent anthropogenic interference with the climate system. Both The Netherlands and the European Union (EU) believe that, given their historic responsibility for the growth in atmospheric concentrations of carbon, the industrialized nations should take the lead in reducing greenhouse gases and should make considerably greater efforts than the developing countries. It is interesting to note that in The Netherlands' Climate Policy Implementation Plan (1999), three packages of measures are worked out: the basic package, the reserve package (if the basic package is not successful enough) and the innovation package (for the period after the first commitment period 2008–2012). Elements of the basic package are energy savings in the traffic sector, CO_2 reductions from energy savings in households and businesses, renewable energy, measures aimed at coal-fired power plants and sequestration in forests, and reductions in emissions of non-CO_2 greenhouse gases. The reserve package consists of an increase in the regulatory energy tax, increased excise duties, N_2O reductions in the chemicals industry and CO_2 storage. In essence for the long-term policy, the innovation package is aimed at instrumental innovation (which would create the possibility of trade in emission and reduction rights) and technological innovation (which would make climate-neutral energy carriers an integral part of all existing policies).

In The Netherlands' Climate Change Policy Implementation Plan part II (2000), The Netherlands' use of the Kyoto instruments in cooperation with foreign countries is worked out. It is in this document that the flexible instruments (JI, ET and CDM) are discussed.

In terms of current policy in The Netherlands, 'long-term policy' refers to the next 20 years, which is definitely not a real long-term approach. One could argue that, from the perspective of the two visions discussed here, the Government of The Netherlands has a very short-term approach!

Working out the legal approach

The most relevant considerations for the legal approach are the binding nature of the law and the criteria for justice under the law. The first question is, 'Who is *responsible* for the problem for which a legal solution needs to be found?' (See also Chapter 5 for the responsibilities of households and Chapter 6 for the responsibilities of local authorities.) From an intersectoral perspective, the following three legal items are crucial:

1. instruments and norms
2. compliance and enforcement
3. public participation and legal protection.

Fundamental legal basics, different fields of law:'Who?'

In the field of law, there are basically two fundamental questions that are essential: What is the binding character of law? Which criteria can be used for valuing justice under the law?

One group of lawyers (the positive legal stream) will answer the first question: via the law itself. Another group (the natural right legal stream) will point out that in essence a part of the answer can also be found in ethics. Those people in the middle will say that the answer can be found in morals, ethics and principles.

In the Introduction, we mentioned two main fields of law: the sectoral and the intersectoral. The sectoral fields of law are also called 'functional fields of law'. For a specific area of policy like energy or the environment, all legal elements are brought together in that specific sectoral field of law. This approach is too narrow for the kind of questions that are relevant here because links have to be made between environmental problems, the energy market and increasing changes in society. We need broader and more general links between these specific fields; therefore, the general, intersectoral fields of law will give more adequate answers. The three main intersectoral fields of law that could be relevant in relation to the problem are (1) international law, (2) national public law and (3) national private law. A brief description of these fields is given below.

International law can be divided into the law of nations and international institutions and the law of regions, such as European law. The bases for the law of nations (and international legal relations) can be divided into the following: treaties (between states), customs, general principles and the binding decisions of international organizations. The bases of European Union law include treaties, regulations, directives and recommendations.

National public law is part of constitutional law, administrative law and criminal law. The main bases of national law include legislation and jurisprudence. In constitutional law, attention is given to the organization of the state on both a centralized and decentralized level, the parliamentary system and the fundamental rights of citizens. Administrative law deals with the organization of administration, the instruments of administration, the norms for using these instruments, the enforcement of the norms and the legal protection for using instruments and enforcing norms. Criminal law is focused on crimes and penalties (sanctions) that can be imposed by a criminal judge (court); there are also relevant aspects of organization, norms and legal protection.

National private law involves contract law and tort law; but there are also relevant norms and legal protection under private law.

Each of these three fields – international law, national public law and national private law – give an answer to the question, 'Who?' In international law, special attention is given to the position of international organizations (in relation to national institutions) and the relationship between states. In national public law, attention is on the position of national public institutions; and in national private law, the focus is on private companies and individual persons. In practice, there is a certain amount of overlap in these fields.

Legal instruments, norms: 'How?' and 'What?'

Each of these fields of law we have mentioned contains legal instruments, and from a general perspective, each can be considered as an instrument itself. But they also contain more concrete instruments, the most important of which - in international and European law - are treaties, decisions of international organizations, EC regulations, EC directives and EC recommendations. In constitutional law the central legal instruments are the constitution, the legislation and the delegated legislation. Administrative law has the following instruments: generally binding rules, other general rules and decisions. Criminal law contains the legislation and judiciary power for imposing principal penalties (both financial and in terms of freedom) and accessory penalties. In private law, the contract is the most common instrument. Some instruments can be seen as primary instruments (generally binding rules or contracts); other instruments, such as instruments of enforcement (administrative sanctions and criminal penalties), are secondary.

International law is focused primarily on maintaining order and peace. However, over the course of time, the aspect of peace has become much more broadly interpreted to include ensuring minimum welfare for all by means of international organizations that have a functional or regional goal. Constitutional law sets norms through several aspects of the rule of law: separate powers, an independent judiciary, protection for human rights. And administrative law is built upon fundamental legal conditions, especially the principles of good governance, public participation and legal protection of citizens. In criminal law, the relevant norm is the legal basis of sanctions, the principle of the innocence of the accused and proof of guilt. In private law, the relevant norms are legal, but the duty of care must also be included.

These legal instruments and norms are essential if the energy system of The Netherlands is to reach the target of an 80 per cent reduction in greenhouse gas emissions by 2050, as discussed in Chapter 2. Specifically, these instruments and norms must be linked to technological change and innovation, which, as mentioned in Chapter 5, is a governance challenge. But because of their role in a greenhouse-neutral society (Chapter 6), local authorities must also pay attention to these instruments and norms.

Compliance, enforcement: 'Who?' and 'How?'

Compliance is the result of implementation and enforcement. In relation to national enforcement, a distinction can be made between judiciary enforcement (enforcement by a judge) and administrative enforcement. Administrative enforcement includes supervision and sanctions by using administrative, criminal or private instruments.

International law is made up of two parts: one is the law of international institutions and one is the law of nations. Basic to the law of nations is the concept that nation states, bound by law, can in principle interpret the law themselves. Enforcement is also to be carried out by the states. Of course, states may also voluntarily accept the judgement of an international court of justice. Enforcement is further developed in cases of supra-national organizations, an example of which is the European Union. In the EU, criminal enforcement should take place at the national level. On the European level, emphasis has been placed more and more on administrative enforcement because an administrative type of law system of institutions and instruments has been developed under the European law. The enforcement of constitutional norms is done by administrative supervisors, democratic control (through the parliament) and the judiciary (civil, criminal and administrative). Enforcement can take place through the administration itself, which is under the judicial control of the administrative judge. The enforcement of private law, in particular, is carried out through the civil judiciary, which has several instruments at its disposal. Instruments of enforcement can – as said before – be seen as secondary instruments.

Enforcement of the norms (including compliance) in relation to the instruments is a precondition for successfully addressing societal resistance or negligence (Chapter 3) and the realization of climate-change options for the long term (Chapter 11). If there is no attention given to enforcement of the norms, instrumental developments could be illusory.

Public participation, legal protection:'Who?' and 'What?'

Environmental protection affects every individual person directly. Guarantees of participation and access to information are fundamental concepts of environmental rights and have been incorporated into law. Access to information is a prerequisite to public participation; the main instrument of public participation is hearings or views submitted in combination with the use of legal instruments. In administrative law, public participation is related to the preparation of procedures for several types of orders, such as decisions. After the orders have been fixed, the process of legal protection comes into effect. When a decision is challenged, this is usually done as a complaint from an interested party against such an order, and the complaint is forwarded to the administrative body that made the order. However, in cases where extensive public participation has taken place, a direct appeal to the judge is possible. Sometimes there is also the possibility of a higher appeal.

The 'instrumentalists' sometimes forget that public participation is essential for reaching the goals of reducing greenhouse gas emissions, not only for households (Chapter 5) and industry (Chapters 4, 7 and 8) but also for the authorities at all levels. Some form of public participation is a prerequisite for reaching these goals.

The process of public participation and legal protection are not well developed in international law; nevertheless, there are international tribunals to address the misuse of certain international rules. Legal protection of fundamental rights is more developed under the European Court on the Protection of Human Rights, as is protection under the European Union Court of Justice. These courts fulfill an additional function to national systems for legal protection.

Legal protection under constitutional law is – apart from political and administrative control – provided by the judiciary. This is also the case with criminal and private law.

Legal dilemmas: 'Who?', 'How?' and 'What?'

Several answers can be given to the question, 'From a legal perspective, who is responsible for the problem for which a legal solution needs to be found?' The right answer essentially depends on the possibility of legal enforcement.

The second question in relation to 'Who?' is, 'On what level – international, regional or national (including the possibilities of decentralization) – do we find the legal entity responsible for enforcement?' Generally speaking, the lower the level, the better the possibility of enforcement. Another consideration is that at a lower level, the legal entity can have more direct contact with citizens.

In relation to the question 'How?', so the question which instrument should be chosen, a difference is made between the primary and the secondary instruments. At the national level, the primary instruments include private (voluntary) agreements (contracts) and public regulations and decisions. An important consideration is that public participation is greater in regard to public regulations and decisions than it is in the case of private agreements. However, the latter sometimes are more effective. In essence, this is the same situation in relation to the secondary enforcement instruments.

For the question, 'What?', referring to the contents of the norms, the combination of general and individual interests is often better guaranteed by the public judiciary than the private judiciary. The reason is that, for instance, the administrative judiciary has experience and has developed special norms 'principles of good governance', which makes its legal protection stronger, in practice. The administrative judiciary is also cheaper and less formal.

The legal aspects of vision A and B

In this section the relevant legal aspects of the visions A and B will be worked out and brought together from the socio-economic description of the two visions. The question is, 'What are the intersectoral legal conditions necessary if substantial breaks in the energy system are to be realized?' To answer this question, we must develop some legal queries based on the instrumental-normative legal approach.

In relation to the legal aspects of these two visions, based on what has been discussed in the previous sections, the following questions have to be answered:

1. Who has the responsibility for taking measures on meeting reduction goals?
2. Which legal instruments are to be used to meet this responsibility?
3. What are the legal norms for using these instruments?

4. How will compliance and enforcement take place?
5. What are the norms for public participation and legal protection?

In answering these questions in relation to the two visions, the legal aspects become clear. There are two extremes: the *passive government* line (vision A) and the *active government* line (vision B). This does not mean that in vision B there is no responsibility for citizens and organizations, nor that in vision A there is full responsibility for citizens and organizations. On the contrary, the question is, 'Who has the lead in relation to the responsibilities for the problem in the society: the government or individual entities (persons or organizations)?' The choice for one of these – active or passive government – will have consequences for the choice of the instruments and norms, compliance and enforcement, and public participation and legal protection.

Below, we will first examine the legal aspects of vision A and then the legal aspects of vision B.

Legal aspects of vision A

There are five questions to be answered in regard to vision A.

The first basic legal question about responsibility is relevant for all the elements that are mentioned in Chapter 2: international, social, economic, energy, environment and the use and allocation of space. The responsibility for most of these activities is in the hands of the individual citizen; the government is responsible only for activities related to infrastructure. As discussed above, the assignment of responsibility also has consequences for the instruments and the norms to be used in enforcement. Most attention will be given to private law and the norms developed in that field of law. The responsibility for compliance is focused on the individual; in relation to enforcement, the judiciary has the task of protecting fundamental rights and imposing criminal sanctions. Because of the private nature of the situations under consideration, there is no significant public participation and legal protection is limited to the protection of fundamental civil rights. In the present Netherlands' Climate Change Policy Implementation Plan part I, a lot of the attention is paid to the responsibility of individual citizens and organizations; there is strict governmental responsibility, mostly limited to financial incentives. Taxes levied in the traffic sector provide examples of this: CO_2 differentiation in the purchase tax on passenger cars and motorcycles and tax measures to limit passenger traffic.

In regard to *international relations* in vision A, there is a global community with a market-oriented approach. This could mean that elements of European law will be globalized in the sense that there will be a global economic market with freedom of goods, persons and services. But there could also be a development along the lines of the World Trade Organization (WTO), which could create tension between regional and global developments. In such a case, the law would have to play a harmonizing role. Now it appears that regional development is more likely because not every country is willing to turn part of its sovereignty over to this global institution. There will not be many directly binding regulations; preference will be given to directives or even to softer *recommendations*. This regional element is already part of The Netherlands' Climate Change Policy Implementation Plan in the form of the EU agreement on fuel-efficient cars and EU regulations on N_2O emissions from catalytic converters on cars.

Compliance will not be a big problem because responsibility is in the hands of individuals, nor will enforcement on the international level be a problem. Criminal enforcement will take place at the national level, as will administrative enforcement, in principle. However, in regard to supervision of observance, administrative enforcement will partly take place, at the international level. Legal protection of fundamental rights will lie in the national courts, the European Court on the Protection of Human Rights and the International Court of Justice.

The *social environment* will involve a small government that only gives very general guidelines. Rapid decision making implies fewer norms for administration and no real need for participation, with the central norms lying more in private law than public norms do. The focus is on individual progress and, therefore, the courts will have the important task of protecting fundamental rights and ensuring against repressive punishment. The focus is on criminal and private enforcement; because of the low government involvement, administrative enforcement is hardly needed.

There will be rapid growth in income and the *economy*, both nationally and internationally. There are no social tasks for the government because most of the social tasks will be taken care of by private organizations. In general, there is less need for public control and subsidies, so there is little development of norms of good governance. The infrastructure – roads and cities – which is traditionally a government responsibility, is growing strongly. Citizens pay for the use of the roads and the facilities provided by the government, as illustrated by the plan in The Netherlands' policy about road pricing. The government has the task to provide suitable conditions for the transport of goods and persons. For this task, administrative law – instruments, norms and enforcement – will be developed, but for other tasks, the government will use private law and then private norms will be relevant. Criminal and administrative enforcement will not really be developed, nor will legal protection be very relevant.

Regarding the *energy and the environment*, there will be a remarkably low price for both. People are not prepared to adapt their behaviour to reduce CO_2 emissions, which implies high technological development and low costs. The consequences for administrative law are the use of financial instruments such as subsidies and taxes (see also the trading instruments discussed in Chapter 9) with less emphasis on mechanisms of command and control. Fiscal mechanisms, such as renewable energy, energy savings in industry and households and energy-efficient appliances, are very strongly developed in The Netherlands' Climate Change Policy Implementation Plan, especially in the field of CO_2 reductions from energy savings in households and businesses. There could also be specific subsidies for the development of alternative sources of energy in which a link could be made with other aspects of the environment such as waste disposal and nuclear power.

The norms are also more individual, and their enforcement should be focused on financial-administrative and criminal sanctions. Private instruments like agreements (covenants) will be very useful in this situation. In general, legal protection will not be very relevant, but it will be important in regard to financial-administrative sanctions.

In regard to the use and allocation of *space*, there will be many changes: more space for people, for industrial complexes, for infrastructure such as roads and airports and much less space for agricultural development. The landscape will have a park-like layout. Norms will have the character of generally binding rules, and there will not be any development of detailed normative permits. This means that there will be a great deal of planning instruments of law, especially in regard to different plans for using or allocating space. Unlike private and criminal enforcement, administrative enforcement will be very important. The administrative court will pay particular attention to legal protection.

Legal approach of vision A

To summarize, the legal approach behind vision A could be called 'a modern liberal state of law'. In such a vision, international law will be especially well developed. It would appear that we will have a world in which a global economic society has been developed. European Union law will be upgraded to 'Global Union law' in some ways, where the regions of the world will be less important. The legal consequences of climate-change policy will in general be partly integrated in individual responsibility and private law and partly in global international law through recommendations.

On the national level, only some classical functions of the state will be important: the protection of fundamental civil rights and against repressive criminal enforcement. In both areas, the judiciary will fill an important function. The protection of social rights will not have priority, neither will administrative law (including public participation and the development of other norms of good governance) be important. Vision A is an individual society in which civil rights are essential and private law is strongly developed.

Legal aspects of vision B

As in the discussion of the legal aspects of vision A, we will also first answer the question, 'Who is responsible for reducing emissions?' in vision B. The answer can be drawn from an analysis of all the relevant considerations: international, social, economic, energy, environment and the use and allocation of space. To meet social needs, the government has an important role to play, which means that the government is in charge of most of the activities. However, it also has the task of developing individual support: individual and social developments are interrelated.

In vision B, there is a variety of *international relations*, regional (and national) developments and differences among regions (and nations), looking for their own identity and trying to be self-sufficient. This means that European (and national Dutch) law will be developed rather than international. It is a situation in which the integration of European and national law in European administrative law will be essential. An example of this in The Netherlands' Climate Change Policy Implementation Plan is the EU agreement on fuel-efficient cars. The consequences are more European legislation and more differentiated European legislation, more enforcement on a European level and legal protection being more the result of cooperation between national and European courts.

In this vision, the implication in regard to the *social environment* is an important role for the administration to protect the general interest, which is the same as individual interests together, along with strong development in the use of instruments of administrative law. In the present Netherlands' Climate Change Policy Implementation Plan, a link is often made between the covenant and environmental permits, although this is not a link which can now be found in the legislation. Another link is the energy regulations in environmental permits. Decentralization on a European and national level will be important, and as a consequence, administrative enforcement will also be more strongly developed along the lines of reparatory sanctions (i.e., sanctions that are focused on the redress of the infringement on the legal system) and punitive sanctions. In such a situation the administrative court will play an important role, and the criminal court will play a more pro-active role in which the preventive aspect of its work predominates.

The growth of the *economy* has consequences for financing the administration, which focuses on more than just the infrastructure. Subsidies will be an important instrument of administrative policy, as we see in The Netherlands' Climate Change Policy Implementation Plan. Administrative enforcement, administrative participation and legal protection will be essential. There will be higher prices for ecological products, stimulated by the administration. Next to subsidies, taxes can also be used to provide environmentally friendly incentives for compliance.

The *energy and environment* aspects seen in the broader context of sustainable development and society will have the result that on regional and local level administrative instruments can be used, especially the command and control instruments. People will accept measures that have a direct effect on their everyday behaviour. This, in combination with the demands of the public transport sector and the needs of the environment, will result in a more highly developed system of administrative instruments of law and mechanisms for enforcement. The administrative court will play an important role.

In regard to the use and allocation of *space*, administrative planning instruments and enforcement will be strongly developed because people will be living in compact, busy cities where living and working are strongly integrated. On a national level, national parks and public areas will become very important. Enforcement of the norms related to the use and allocation of space will be very important. Consequently, the administrative court will play the leading role, not only in relation to conflicts between citizens and the government about permits but also in relation to enforcement.

The legal approach of vision B

The legal approach behind vision B can be called *a* 'modern social state of law'. In this vision, regional law - especially European law - rather than international law will be strongly developed. The consequence will be the development of bureaucracy on the European level. In addition, a strong government will be necessary at the national level. The central questions will deal with the protection of social rights on the one hand and the tasks of government on the other.

Enforcement will not only be criminal, but also administrative. Private law will only fulfil a 'rest function' in relation to enforcement (i.e., private law enforcement will only be used when criminal or administrative enforcement cannot be applied). Public participation and the development of other norms of good governance will be important.

Comparing the legal aspects: two legal approaches

In comparing the legal aspects of the two visions, it should first be made clear which legal aspects provide the basis for the comparison. It seems fruitful to do that in terms of fields of law in relation to the elements of responsibility, instruments and norms, enforcement and public participation and judicial review.

The discussion of the two visions has made it clear that several legal fields are relevant: international law (international and regional), national public law (constitutional, administrative and criminal law) and national private law (contract and tort law). This means that we have a multi-disciplinary legal approach in which it should be clear that the importance of each field of law and its relevant aspects are different for the two visions.

In vision A, the emphasis is on international cooperation, whereas in vision B the emphasis is on the regional level and self-sufficiency. This means that international law will be developed more in vision A, and regional law will be developed more in vision B. In case of The Netherlands in vision B, this will be European law. Self-sufficiency on a regional level should also have a place in European law. In both cases, the question is how national (public and private) law should be developed. In vision A, there is strong development of national private law, especially contract law and criminal law; in vision B, there is strong development of national public administrative law, with an additional role for national private law.

Vision A predicts a competitive, individual society with a stronger preference for individual interests. In terms of law, this means that the protection of fundamental civil rights will be important, and strong development of criminal law and private law will take place. Vision B gives us a more socialized approach with promotion of general interest, implying development of detailed administrative laws with aspects of legal instruments, enforcement and legal protection.

We can draw the following conclusions in regard to the consequences for climate policy based on these two models. In vision A, there is a modern liberal approach, the liberal state of law. In vision B, there is a social state of law. In vision A, there is primarily strong development of soft international regulations on climate change, in which member states (parties) do not have a strong position and there are fewer regulations and the international free economic market is decisive. Canada and Australia seem to have this approach in relation to climate change policy. In vision B, there is an active government line with strong regulations on the regional (European) and national level, and the government takes a strong position on protecting the general interest in regard to climate change. The instruments of administrative law, norms, participation and enforcement will be strongly developed at both levels. Social responsibility will be channelled through public participation. Private law will have a rest function. This approach is seen more in the European Union and its member states.

Both visions present extremes and are therefore not completely feasible in reality. That is also clear when one realizes that realistically, in any situation, there will always be a combination of responsibilities, instruments, norms, enforcement, public participation and legal protection.

But keeping these remarks in mind, we discuss below some extreme theoretical consequences of the legal aspects of the two visions.

First, some theoretical legal consequences of an energy system in vision A: in that vision there is a liberal world energy market; market forces and cost minimization will tend towards larger-scale production capacity instead of a decentralized supply system. On the national level, governmental instruments for commanding and controlling compliance will not be needed. Furthermore, no special attention will have to be paid to the traditional aspects of enforcement. If there is a need for control measures to be taken, preference will be given to trading in emission rights because this is an explicit market instrument that minimizes costs. There is no need for public participation; legal protection is fixed on protecting fundamental rights.

Second, some theoretical legal consequences of an energy system in vision B: there are far-reaching measures to ensure energy efficiency throughout the entire economy, and there is a preference for adapting the existing infrastructure instead of designing and installing completely new energy systems. Renewable energy sources are strongly supported; nuclear energy is mainly left out of the Dutch energy supply mix. There is a strong need for regional, European and national measures, and special attention will have to be paid to enforcement. Because of the rejection of the market as a controlling force, only command-and-control mechanisms are needed. Public participation is essential and legal protection deals not only with protecting fundamental rights but also with the protection of social rights.

The consequence for the climate change policy is that there is a need for a combination of the modern liberal and social legal approaches. The question is now: how is the relation to the present climate change policy of The Netherlands? That will be worked out in the next section.

Confronting legal approaches with contemporary policy

This assessment will be based on two sources: first, the opinion of the Dutch government, which can be found in The Netherlands' Climate Change Policy Implementation Plan; second, the normative-instrumental approach developed here.

The long-term approach of the Minister of Environment

The Minister bases his prospects for the long term (20 years) on five developments: (1) ultimately greenhouse gas emissions will have to be drastically reduced to prevent anthropogenic interference with the climate system (80 per cent for the Annex I countries is not an improbable ultimate goal), (2) the long-term significance of the flexible Kyoto instruments is uncertain, (3) the role of the non-CO_2 greenhouse gases will decline, (4) the reserves of fossil fuels will not be exhausted within the foreseeable future, and (5) energy prices may remain low.

The Minister agrees with the conclusion that reducing CO_2 emissions within The Netherlands will have to play an increasingly important part in policy-making after the first commitment period. Both the potential and price of project-specific reductions via the Kyoto Mechanisms Joint Implementation and the Clean Development Mechanism are uncertain. On the larger scale, CO_2 reductions will only be possible if, in the longer term, the transition is made to an energy system that emits much less CO_2.

At the request of The Netherlands' government, a study has been made on long-term reduction possibilities for The Netherlands (up to the year 2020). An enormous policy effort would be needed to reduce emissions. Two policy variants were used to study the consequences of such an effort in an attempt to define the outer limits of the playing field. The variants differ in the role that the government plays. In one, the regulatory variant, government activity directs the effort, making use of policy instruments such as regulations combined with financial incentives. In the other, the market variant, the government limits itself to creating optimal conditions and leaves it to the market to come up with the necessary reductions in a cost-effective way. In this second variant, less use is made of regulations and subsidies and more use is made of levies and tradable permits.

The study shows that CO_2 emissions can be reduced by 95 Mtonnes in both variants at a net cost of about 4.5 billion Euros per year. The macro-economic impacts are negative, but slight. The economy grows – in this study – as a result of the reduction policy by 121 per cent to 123 per cent over a 25-year period, rather than 125 per cent. Growth in employment is also marginally lower, 30 per cent rather than 31 per cent. Emissions are reduced in slightly different ways in the two variants. The share of energy conservation is clearly larger in the regulatory variant (government regulations require non-profitable measures to be taken), while a lot of biomass is imported in the market variant.

Preparations have been made for the long term through technical and instrumental innovations. The Dutch government has agreed with the necessity of policy instruments that make optimal use of the self-directing capacity of society and the economy. General market-oriented instruments, such as levies and tradable emissions and reductions, are most suitable for this purpose. However, an emission ceiling for energy-intensive, export-oriented sectors is the most sensible and feasible approach if international trade is to be possible on a large enough scale through a system of tradable emissions for businesses. A national ceiling on emissions from these sectors would place their competitive position at risk if other countries do not impose such a ceiling. In this regard, tradable emissions would share the limitations of energy taxes as a national policy instrument. However, command-and-control instruments are more costly, compared to the emission ceiling of tradable permits, which ensures effectiveness provided that enforcement functions well. Therefore, three lines will be followed: (1) making a start with a domestic system of tradable emissions for the so-called *sheltered* sectors and households, (2) continuing existing policies for the sectors not participating in the system of tradable emissions through long-term agreements, benchmarking and regulations, (3) bringing energy-intensive, export-oriented sectors under an international system of tradable emissions as rapidly as possible. (See also Chapter 9.)

The Minister's approach, the two legal visions and an alternative

In terms of legal visions, the approach of the Minister of the Environment in The Netherlands' Climate Change Policy Implementation Plan can be qualified as a modern liberal approach. Of course, it is not a purely modern liberal approach because it also includes some clear tasks for the government; therefore, there are also some elements of the social legal approach in it.

Another important point is that the Minister's approach for the long term (the next 20 years) is not very well developed and there will have to be more done at the national level. The term should also be longer than 20 years.

An alternative for the Minister's approach could be the following. Of course, it is good to have a package of measures for the future, but 20 years is a very short time in relation to the kinds of change that have to take place. So the question is whether the Minister's policy plan is realistic. The general impression is that the outcome is not realistic insofar as the expected outcome is concerned. Looking at the current situation and at current social developments, the question is whether the goals of the policy will ever be reached. It is therefore interesting to look at the legal approach behind the plan. The plan gives the impression that it is more a modern liberal approach than a social legal approach.

After evaluating the policy plan, it appears that there is a need for stronger policies on climate change. Our suggestion is to bring more elements of the social legal approach into the plan. This means European regulation; more responsibility on the part of the government, including public participation; more command-and-control instruments; more attention to enforcement and legal protection of social rights.

Conclusions and recommendations

Vision A has a modern liberal legal approach; the legal approach behind vision B is a social legal one. In vision A there is a passive government and in vision B there is an active government. This does not mean that in vision A the solution of the problem is provided by citizens and organizations and in vision B by the government. In essence, the question is the same as in film production: Who is responsible for direction? Is it especially an active government or is it citizens and organizations in combination with a passive government?

In a system of permit trading, the government has a passive function for directing the process. In addition, the system is not well developed in relation to either fundamental legal norms or enforcement. Permit trading is an instrument that is related to the modern liberal legal approach. The choice for a more active government has consequences for the choice of the instruments and norms, for compliance and enforcement and for public participation and legal protection.

This means that in the active approach, there is a preference for public law instruments (and less for private instruments), while in the passive approach, priority is given to private law instruments (and less to public instruments). In regard to public instruments, the norms are particularly related to the general interest (more related to social considerations), while the private instruments are more related to individual interests. In regard to the private instruments, compliance and enforcement are also primarily the responsibility of individuals. Compliance and enforcement in regard to the public instruments are the responsibility of the government. Also, public participation and legal protection are more developed in relation to public instruments in legislation and are more related to the general interest. The private instruments reflect more a question of individual interest.

Therefore, in theory, and thinking in terms of the two visions discussed here, from the legal point of view for the future, the choice should be an active government role, as described in vision B.

In regard to this theoretical choice, the following solutions can be given for these dilemmas. The responsibility for direction lies at the level of the government, partly at the European level and partly at the national level (including provincial and local): European regulations and directives based on public participation and implementation by the member states. These regulations would contain instruments and norms that would stimulate the activities of citizens and organizations to reduce CO_2 emissions directly. A well-developed mix of administrative and criminal enforcement at both levels would be an essential part of the chain. This trend can be seen better at the international (EU) level than at the national level.

The legal consequences for climate change policy depend on the theoretical legal vision chosen, but in practice the outcome will be different as there are other influences that make the situation more complicated. In practice, a well-balanced combination of the modern liberal and the social legal approaches will give the best legal results. Our feeling is that the active government line – the basis of the social legal approach – is underdeveloped and that after some time the outcome will be less positive than is currently predicted. In the future, there will be a need for a plan based more on a combination of the modern liberal and social legal approaches. Only then will the radical changes needed in the energy system in The Netherlands by the year 2050 be possible in a legally acceptable way.

Bibliography

Addink, G.H., Norms and enforcement of the Climate Change Convention, *Climate Change and the Future of Mankind*. Pre-Cop 3 International Symposium on Legal Strategies to Prevent Climate Change, Tokyo, 1997.

Addink, G.H., Implementation and enforcement of the Kyoto Protocol after Buenos Aires November 1998, *Jahrbuch des Umwelt- und Technikrechts*, Berlin: Erich Schmidt, 1999.

Addink, G.H., Joint Working Group Compliance on the Kyoto Protocol: an overview of suggestions on compliance, in: *El Protocol de Kyoto*, San Jose (C.R.): Instituto de Investigaciones Juridicas, 2000.

Arts B., New Arrangements in Climate Policy, *Change* 52, pp. 1–3, 2000.

Baker, S., *Environmental Governance in the EU*. Quebec: IPSA Paper, 2000.

Cozijnsen, C.J.H. and G.H. Addink, The Kyoto Protocol under the Climate Convention: commitments and compliance, *Change* 43, August/September 1998, pp. 5–8.

Cozijnsen, C.J.H. and G.H. Addink, Het Kyoto-Protocol onder het Klimaatverdrag: over de inhoud, de uitvoering en de handhaving van afspraken, *Milieu en Recht*, 6, 1998, pp. 152–159.

Glasbergen, P. (ed.), *Co-operative Environmental Governance: Public-Private Agreements as a Policy Strategy*. Dordrecht: Kluwer Academic Publ, 1998.

Golub, J. (ed.), *New Instruments for Environmental Policy in the EU*. London: Routledge, 1998.

Gupta, J., C. Jepma and K. Blok, International climate change policy: coping with differentiation. *Milieu* 13(5), 1998, pp. 264–274.

Kohler-Koch, B., and R. Eising (eds.), *The Transformation of Governance in the European Union*. London: Routledge, 1999.

Kolk, A., Multinationale ondernemingen en internationaal klimaatbeleid. *Milieu*, 14(4), 1999, pp. 181–91.

Lefevere, J. and F. Yamin, *The EC as a Party to the FCCC/KP, An examination of the EC competence*, London: Field, 1999.

Oberthür, S. and H.E. Ott, *The Kyoto Protocol, International Climate Change Policy for the 21st Century*, Berlin/Heidelberg/New York: Springer, 1999.

OECD, *Ensuring Compliance with a Global Climate Change Agreement*, OECD Information Paper, Paris, 1998.

Ott, H.E., Outline of EU Climate Policy in the FCCC? Explaining the EU-bubble, *Climate Change and the Future of Mankind*. Pre-Cop 3 International Symposium on Legal Strategies to Prevent Climate Change, Tokyo, 1997.

Rolfe, C., *Kyoto Protocol to the UNFCCC: a guide to the Protocol and Analysis of its Effectiveness*, West Coast Environmental Law Association, 1998.

UNEP, *Report of the Working Group on Compliance and Enforcement of Environmental Conventions*, Nairobi: UNEP, 2000.

Van der Jagt, J.A.E., Elaborating an international compliance regime under the Kyoto Protocol, in: E. Van Ierland, J. Gupta and M.T.J. Kok (eds.) *Options for international climate policy*, Cheltenham, E.Elgar, 2002 (in press).

Werksman, J., *Responding to Non-Compliance under the Climate Change Regime*, OECD Information Paper, Paris, 1998.

Chapter 11

Climate OptiOns for the Long term – COOL: Stakeholders' views on 80 per cent emission reduction

Matthijs Hisschemöller, Marleen van de Kerkhof, Marcel Kok and Rob Folkert

In this chapter...

... stakeholder perspectives on drastic emission reduction are presented. The previous chapters analysed the transition towards a climate-neutral society from different academic focuses. Since substantial transformations will depend heavily on the commitment and co-operation of relevant societal actors, it is vital to consider stakeholders' perspectives on this issue. This chapter addresses the views of different stakeholders on long-term strategies to realize drastic greenhouse gas emission reductions. These insights were generated during an 18-month stakeholder dialogue in The Netherlands. The future images described in Chapter 2 served as a starting point for this stakeholder dialogue. The stakeholders identified opportunities and threats to realize reductions up to 80 per cent in 2050 compared to 1990 levels. These are linked to the dilemmas presented in Chapter 3 of this book. The approach used in this project illustrates the type of work advocated by many of the authors in previous chapters when underlining the need for stakeholder participation, social dialogue and co-production.

Introduction

Stakeholders have a decisive role in achieving emission reductions. Based on the National Dialogue of the COOL project (Climate OptiOns for the Long term) (Box 11.1), this chapter presents insights and recommendations with respect to how The Netherlands can prepare itself to make the transition towards a climate-neutral society. The two visions of the energy system in The Netherlands described in Chapter 2 formed the starting point for the stakeholder dialogue.

The dialogue was conducted in different groups, each concentrating on one of four sectors: housing, industry & energy, agriculture & food, and traffic & transport. The participants in the dialogue, persons with a record of service in, and broad knowledge of, the sector concerned, differed in disciplinary and professional background. The dialogue included stakeholders with different backgrounds such as: government (local and national), the business community (NUON, Siemens, DSM, CORUS, SHELL, Akzo Nobel, an ICT company, European Business Council), banks (National Investment Bank, Rabobank), non-governmental organizations (Greenpeace, Church & World, World Wildlife Fund, The Netherlands Trade Unions Federation (FNV), Global Action Plan (GAP), Netherlands Association for Countrywomen), and umbrella organizations (the

Box 11.1

The National Dialogue as a process

The National Dialogue was part of the Dutch COOL project. Besides the dialogue at the national level, COOL also included a dialogue at the European and global levels.

The National Dialogue was co-ordinated by a project team and a scientific support team. The project team's main responsibility was to organize and facilitate the dialogue and to report its results. The scientific support team provided the dialogue groups with 'state of the art' scientific information on options to reduce greenhouse gases in The Netherlands. The members of the scientific support team were also involved in the project team. The project was carried out in three phases: a design phase, the actual dialogue phase and a reporting phase.

During the *design* phase the project team interviewed about 100 persons from the four involved sectors on the dialogue's scope and focus and the possible composition of the dialogue groups. These interviews contributed a great deal to the ultimate shape of the actual dialogue. For each dialogue group, chairpersons were contacted. About 60 persons accepted the invitation to participate in the dialogue on an individual basis. In the meantime, the scientific support team developed Future Images (see Chapter 2). The dialogue participants received two notes for approval, the first addressing the dialogue's Scope and Rules of the Game, the second dealing with the dialogue's Process and Schedule.

The actual *dialogue* phase was divided into three steps:
1) First, the dialogue groups discussed the Future Images presented by the scientific support team (Faaij et al., 1999). Inspired by these images, they developed two images for their sector, both based on the assumption of -80 per cent greenhouse gas emissions.

These images constituted the point of departure for the analysis of options in the second stage of the dialogue.

2) The groups then selected options for emission reduction for further analysis (see Box 11.2). They analysed the options using the backcasting technique. This meant that the dialogue group took the implementation of a specific option as a starting point. For a period of 50 years from now, the chances and obstacles for the option were identified and the major problems solved. Finally, the results of the backcasting exercise were visually presented on a time sheet.

3) Finally, the groups compared the outcomes of the separate backcastings in order to identify criteria for long-term climate policy (see Box 11.5). Then, they identified clusters of options that, to their point of view, meet their criteria, and identified actors and policy instruments which they considered vital for implementation.

Each dialogue group has met six times between November 1999 and March 2001. Representatives of each group also met at two joint workshops, where they exchanged interim results and discussed their conclusions.

In the *reporting* phase, the findings for each sector were laid down in strategic visions, which are published in the end report of the National Dialogue (Hisschemöller and Van de Kerkhof (eds.), 2001); also an integrated report on the National Dialogue was written (Hisschemöller, 2001). The various findings and reports were reviewed by the participants prior to publication. The responsibility for this chapter rests entirely with its authors. For more information about COOL, see www.nop.nl/cool

National Harbor Organization, Nederland Distributieland, the organization of the greenhouse industry). All the stakeholders participated in the dialogue on an individual basis.

The groups explored options for long-term climate policy. They focused on what would be necessary to reduce emissions in The Netherlands in 2050 by up to 80 per cent compared to 1990 levels. In other words, the participants did not concern themselves with whether an 80 per cent emission reduction is a *desirable* target for climate policy.

Box 11.2
Options analysed in the National Dialogue groups

- Biomass (Industry, Agriculture, Transport)
- Chain optimization of wood (consumption) (Agriculture)
- Combined heat and power (Industry)
- CO_2-neutral greenhouse (Agriculture)
- CO_2 removal and storage (Industry)
- Curb demand for transportation through behavioural change (Transport)
- Energy efficiency in industry (Industry)
- Fuel cell (Transport) (also see under Hydrogen)
- Heat pump (Housing)
- Hydrogen economy (Industry, Transport)
- Measures to reduce emissions through land management (Agriculture)

- Measures to reduce emissions from manure and fermentation (Agriculture)
- Micro-combined heat and power (Housing)
- Modal shift from private to public transport (Transport)
- Modal shift from air to train (Transport)
- Modal shift from road to water (Transport)
- Passive solar (Housing)
- Replacement rate of buildings (Housing)
- Sinks (Agriculture)
- Solar PV (Housing, Industry)
- Underground transport (Transport)
- Wind (Housing, Industry)

The next section of this chapter briefly summarizes the routes towards an 80 per cent reduction of greenhouse gas emissions in 2050 based on the strategic visions of the four sector groups. The remainder of the chapter is devoted to a summary and analysis of the findings and conclusions of the National Dialogue. Where possible, reference is made to the dilemmas identified in Chapter 3. After that the obstacles and opportunities that were identified are described and we go into the criteria which, in the dialogue groups' judgement, should guide climate policy. Next, we discuss some pivotal issues related to crucial options for long-term climate policy and the role of government in long-term climate policy. To conclude, recommendations are formulated, dealing mainly with how The Netherlands can prepare for the longer term during the coming 5-10 years. The last section presents the most important lessons from the National Dialogue as a process. Together, these recommendations make up the ingredients that are regarded as necessary for the development of a long-term climate strategy.

Trajectories for drastic greenhouse gas emission reduction in sectors

As part of the National Dialogue, trajectories (packages of options) were formulated for each sector. Most of these are complementary, but some conflict with each other. Box 11.2 surveys the options that were considered by the groups as they developed their visions. All the options analysed could potentially make a valuable contribution to a drastic reduction of emissions in The Netherlands.

Housing

For the housing sector, two complementary trajectories towards a solution were proposed: one for existing buildings and one for new construction. It was assumed that what holds for the reduction potential in the housing sector holds comparably for the utility sector.

The package for **Existing Buildings** is made up as follows (in order of preference):
- dwelling insulation (roof, crawl space and house front)
- sustainable energy applications (solar boiler, solar PV, wind)
- low-calorific heating; heat pump combined with micro-combined heat and power (CHP) (as successor to the high-efficiency boiler).

The package for **New Buildings** includes (in order of preference):
- integrated design with optimal orientation to the sun and use of daylight
- optimal thermal insulation
- balanced ventilation with recovery of heat
- sustainable energy applications (solar boiler, solar PV, wind)
- low-calorific heating: heat pump combined with micro-CHP (as successor to the high-efficiency boiler).

The group expressed a strong preference for sustainable energy options that can be applied to individual buildings. It also sees a potential for wind energy. Consumers are expected to prefer renewables and even be willing to pay more for

---

Box 11.3

CO_2-emission reductions for two trajectories in the sector Housing

A calculation for houses shows the following results for 2050:

- Under **autonomous developments** (i.e. without addition of options), CO_2 reduction through improved efficiency in buildings and power stations will exceed the increase of CO_2 emissions as a consequence of the increase of the total building stock. CO_2 emissions will decline by -10 to -25 per cent in 2050 compared to 1990 levels.
- If **additional reduction options** *inside* the dwelling are used to their maximum potential, CO_2 emissions will decline further to -60 to -70 per cent compared to 1990 levels.
- If **'clean' electricity** can also be supplied by the grid, a further reduction will be possible up to -80 to -90 per cent compared to 1990 levels.
- Hence, 80 per cent reduction does appear within reach.

them if necessary. CO_2 removal and storage should *only* be applied if renewables prove insufficient to reduce emissions by 80 per cent in 2050. By then, the fossil-based share of the energy supply will also become CO_2-neutral.

Depending on the assumptions related to the replacement rate of houses and utility buildings, there is the prospect of emission reduction up to 80–90 per cent (Box 11.3). As the pace of replacement accelerates, it will be possible to realize further reductions (emissions related to demolition and reconstruction of buildings included). The question remains whether sufficient sustainable electricity (wind, solar) can be produced from the grid, or whether clean fossil fuels must also be used (and consequently CO_2 storage must be applied).

Industry & Energy

The first observation with respect to the industrial sector is that a major improvement in efficiency is indispensable if emissions are to be reduced by 80 per cent by 2050. A lot is expected from combined heat and power, especially in the short and medium term. The second observation is that all supply options that really matter - biomass, CO_2 removal and storage, and renewables (wind, solar) - are, for different reasons, controversial. However, if an 80 per cent reduction is to be realized by 2050, the variant most likely to succeed is for all available options to be applied, in other words: increased energy efficiency, renewables and CO_2 removal and storage.

Three conflicting trajectories were identified for the sector.

- In the trajectory **Clean Fossil**, 80 per cent of the energy carriers come from fossil sources, natural gas in particular. The focus here is on developing a hydrogen infrastructure, CO_2 storage, biomass and CHP. It is projected that efficiency improvement will be modest (35 per cent compared with the 1990 level, or 0.75 per cent per annum).
- In the trajectory **Sustainable Energy system**, about 70 per cent of the energy carriers come from renewable sources, imported biomass in particular. There is also a focus on solar and wind. A high level of efficiency improvement is projected (50 per cent compared with the 1990 level, or 1 per cent per annum). In this trajectory, no effort is made to develop the clean fossil package, CO_2 storage, hydrogen and CHP.
- The third trajectory is a **Hybrid**, which combines elements of the other two except for biomass. It includes solar and wind but also CO_2 removal and storage, hydrogen and CHP. For this trajectory, a low rate of efficiency improvement is projected (20 per cent compared with the 1990 level, or 0.4 per cent per annum).

Depending on the assumptions concerning sector growth, efficiency improvement and the contribution from CO_2-neutral supply options, the industry & energy sector might realize an overall reduction up to 50–75 per cent by 2050 compared with 1990 levels (Table 11.1). Additional reductions could be realized especially through the use of CO_2-neutral feedstocks (biomass in particular) in the fertilizer and chemical industries. If all these options are available, it would (theoretically) be possible to achieve a reduction of almost 100 per cent for the industry & energy sector.

Table 11.1 *Emission reductions of CO_2 for three trajectories in the sector Industry & Energy*

Scenario & supply mix	Clean Fossil		Sustainable Energy system		Hybrid	
	Emission	Reduction	Emission	Reduction	Emission	Reduction
	Mt CO_2		Mt CO_2		Mt CO_2	
Low growth	16	75%	15	76%	18	72%
High growth	26	60%	26	60%	29	55%
High growth, no materials efficiency	30	54%	31	52%	34	48%

For an understanding of these figures: The 1990 CO_2-emissions for Dutch industry were about 64 Mton

Agriculture & Food

In the sector agriculture, policies should be developed and implemented in a European context. The group suggested three complementary trajectories.

- **Measures targeted at primary production**: CO_2-neutral greenhouses, closed stables where animals have sufficient space, organic instead of artificial fertilizers and a huge improvement in the efficiency of fertilizer use.
- **Energy production and chain optimization of wood consumption**. Energy production is based on the available biomass and on-shore windmills. Optimization of wood consumption assumes that wood is primarily used for high-end applications, especially construction, and only in the last instance as fuel.
- It is believed that **interventions in the food chain** will also significantly reduce emissions, but these were not analysed in the backcasting.

In addition to these trajectories, there are opportunities to make use of sinks to contribute to CO_2 reductions. Options in this respect are conversion of arable land to pasture, new afforestation, less deep ploughing and especially increasing the groundwater level in the peat pastures (this prevents peat oxidation and stimulates the formation of new peat). However, there are great uncertainties in the net results.

The range in the figures for the potential reduction of CO_2 emissions presented below is due to differences in the projections for the volume of primary production in 2050.

- Implementing measures related to primary production renders an emission reduction of 12-18Mt CO_2 eq. This yields 60-80 per cent reduction in the remaining emissions (5-10 Mt CO_2 eq.) compared to the 1990 level.
- Energy production and optimization of wood consumption yields an additional reduction in the order of 6-8.5 Mt CO_2 eq. (this is 25-35 per cent of the overall emissions from the sector in 1990).
- Sinks could generate a further reduction of 7-9 Mt CO_2 eq. (also 25-35 per cent of the sector's emissions in 1990).

This means that, in total, the sector could reduce its own emissions by 100-150 per cent in 2050 (150 per cent means that the sector reduces emissions in other sectors as well) (Table 11.2). This figure does not include the results from interventions in the food chain.

Table 11.2 *Emission reduction of CO_2 for the sector Agriculture*

	Reduction potential*) (Mton CO_2 eq.)
Reduction primary production	
▪ CO_2-neutral greenhouse	9-14
▪ Closed stables	2-3
▪ Emissions per cow	0.5-1.5
▪ Organic fertilizers	Uncertain
▪ Highly efficient use of fertilizers	1
Total primary production	*12-18*
Sustainable energy and materials NL	
▪ Utilization of agri residuals	0.5-1
▪ Manure fermentation	0.5-1
▪ Combustion of (dry) manure	1
▪ Bio-energy production NL	1
▪ Wind farms on shore	1-1.5
Total sustainable energy and materials NL	*4-6*
Sustainable energy sources and materials from abroad	
▪ Chain optimization of wood consumption	2
▪ Bio-energy production imported (400-500 PJ)	38
Total sustainable energy sources from abroad	*40*
Sinks	
▪ Increase groundwater level in peat pastures (450 kha)	?5-7**
▪ New forest (350 kha)	?1
▪ Forest and land management	?1
Total sinks	*?7-9*

*) The reduction potential of the options cannot simply be added because of overlap.
**) ? means that the net effect is uncertain

Most remarkable is the sector's possible contribution to reductions in other sectors. If the measures are also applied outside The Netherlands, a shift from artificial to organic fertilizer in combination with efficient use of fertilizer will probably render a reduction in the order of 8-11 Mt CO_2 equivalents (N_2O and CO_2). Cement production will be affected by optimization of the wood consumption chain since the use of concrete in construction will be reduced. In total, agriculture could contribute to a reduction of 14-19 Mt CO_2 eq. in other sectors.

Traffic & Transport

A general observation made about this sector is that it cannot influence developments leading to an increased demand for transport in such a way as to curb the demand.

Four trajectories towards an 80 per cent reduction in emissions by 2050 were formulated:

▪ The trajectory **CO_2-neutral transport fuels** includes the development and use of bio-fuels and / or clean fossil fuels for transport combined with CO_2 removal and storage.

Table 11.3 *Energy use figures for Transport, translated into CO$_2$ figures*

CO$_2$ emissions in Mton	
1990 level	28
Trend 2050	60-75
Total Effect trajectories 1-3 compared to trend	-25
Technical solutions	-15
Reduction of demand	-5
Modal shifts	-5
Results for 2050 without CO$_2$-neutral fuels	35-45 (+20-50%)

- The trajectory **Technological solutions to curb energy demand for vehicles** includes the development and use of fuel-efficient vehicles whereby less energy will be required to meet total demand for transport.
- The trajectory **Curbing demand for transport** includes changing the behaviour of individuals and more efficient commodity transport thanks to ICT, which leads to fewer 'empty' kilometres being driven.
- The fourth trajectory is **Modal shifts**. This includes shifts from road to water, from air to rail and from private to public.

Major breakthroughs for addressing the climate issue are expected from the first and the second trajectories in particular. However, modal shifts toward low-energy / CO$_2$-efficient modes of transportation and reducing the overall demand for transportation are also felt to be necessary. This claim is warranted by two observations. First, the considerable growth of the sector that is expected in the coming decades is likely to put a heavy strain on the availability of CO$_2$-neutral transport fuels. It is therefore assumed that there may be limits to the availability of CO$_2$-neutral fuels (e.g. from biomass and renewables). Secondly, modal shifts appear relevant in the context of issues other than climate change, especially those that relate to scarcity of land and congestion.

Calculations have been made with respect to the possible development of energy use in the various trajectories. It emerged that despite a package of powerful measures, energy use by the sector will have increased by 20–50 per cent in 2050. This figure clearly shows that the use of CO$_2$-neutral fuels in combination with other technological solutions, such as the fuel cell, will be indispensable for realizing far-reaching emission reductions (Table 11.3). Without this option, the realization of any reductions becomes highly uncertain as the growth of the sector exceeds the impact of efficiency improvements. Therefore, the market penetration of CO$_2$-neutral fuels will ultimately largely determine whether the sector realizes greenhouse gas reductions. Theoretically, in the event of complete market penetration, reductions up to 100 per cent will be a possibility.

The outcome of the National Dialogue indicates that an emission reduction of 80 per cent in 2050 would be possible in The Netherlands. Table 11.4 gives a summary of the conclusions of the four dialogue groups in terms of trajectories and reduction potential.

In most cases, the sectoral trajectories are complementary, but some compo-

Table 11.4 *Summary of findings for four sectors in 2050*

Sector	Housing	Industry & Energy	Agriculture & Food	Traffic & Transport
Trajectories	Existing buildings: sustainable New buildings: sustainable	Clean Fossil: CO_2 storage, H_2, efficiency Sustainable Energy System: Biomass, wind, solar, high efficiency Hybrid: CO_2, H_2, sustainable	Primary sector Energy and materials Sinks	Draw back of demand for transportation Efficiency - Modal shifts Clean fuels
Emission reduction	80–90%	50-100%	100-150%	??-100%

nents are mutually conflicting. This is especially true for the Industry group, which formulated three conflicting trajectories: clean fossil, sustainable energy system and hybrid.

In some cases it is possible that trajectories for different sectors may be mutually exclusive. It is questionable, for instance, whether the 'clean fossil' trajectory for the industry & energy sector is in fact capable of satisfying the demand for sustainable energy from the housing sector. It is also questionable whether the demand for biomass can be satisfied in all cases. After all, both the traffic & transport sector and the energy sector put claims on imported biomass for a variety of purposes, for example to satisfy domestic and industrial demand for energy and to meet the demand for clean fuels. An important point is that different trajectories probably make different demands on an expensive infrastructure. A trajectory towards a sustainable energy system presupposes costly investments that are very different from the investments required by a clean fossil scenario. The dialogue observed that such inconsistencies do exist but did not address them further.

Obstacles, opportunities and criteria for long-term climate policy

The findings of the four groups in the National Dialogue are set out side-by-side in Table 11.4. Based on these findings, a considerable number of dialogue participants concluded that, given a number of preconditions - some of which are outside the control of The Netherlands itself - drastic reduction of emissions up to as much as 80 per cent in 2050 are conceivable, but there are doubts about its practical feasibility. Even with a time-frame for implementation as long as 50 years, all trajectories meet with considerable barriers. In some cases, there are grave doubts about whether an option can be realized, such as cutting the demand for mobility by changing behaviour. At the same time, there are opportunities which government and other parties in society can exploit. Box 11.4 shows the obstacles and opportunities that were identified.

> **Box 11.4**
> ## Types of obstacles and opportunities found
>
> **Obstacles**
> - Technology development
> - Vested interests in sector
> - Costs (compared to other options)
> - Public acceptance (a variety of image and acceptance issues)
> - National government (internal co-ordination and enforcement)
> - European Union (especially co-ordination)
> - Infrastructure
> - Scarcity of space
>
> **Opportunities**
> - Technology development
> - Public acceptance / Image
> - Fit with trends in sector
> - Efficient use of scarce space
> - National government

In general, problems that are administrative in nature (especially those that demand a European approach) and the lack of public acceptance are regarded as the primary risk factors. The retreat of government and the liberalization of the energy market appear to conflict with the implementation of a large number of climate options that demand an active government role in many areas. The most important opportunities lie in the area of technological development and, once again, in the realm of public acceptance. Climate policy in The Netherlands could lead to greater domestic comfort, less traffic noise and a pleasant landscape, and it need not have a negative effect on matters to which people are attached, such as their income and mobility. The dialogue thus envisions a possible approach that is directed towards co-benefits (Dilemma 6: 'Transition through co-benefits' versus 'The inevitability of a special climate change approach').

It will be impossible to set all the options in train simultaneously, however. Even over a period as long as 50 years, choices will inevitably have to be made at appropriate moments. This becomes clear if we look at the criteria that the dialogue groups propose to guide the development of long-term climate policy, which are set out in Box 11.5. The main conclusion of the dialogue is that climate policy must be directed at the advancement of options that fulfil criteria of *climate effectiveness, sustainability, cost effectiveness, and societal support*. However, there is sometimes a tension between these criteria, and this is where opinions diverge.

There are differences in opinion and in expectations about what may be feasible and socially acceptable given the current state of technology. It is generally assumed that major improvements in technology will occur in many areas. Some can imagine a drastic cut in emissions without radical innovations. In that case, it would appear obvious to conclude that the most serious obstacles are to be found within government (political will, resolve, consistency, etc.). Other important obstacles relate to the private sector or consumers (acceptance, dilemma 4: 'demystification' versus 'need for an increased awareness of climate change impacts'). Such a conclusion would appear to accord with that of the IPCC in its Third Assessment Report: sizeable reductions in greenhouse gas emissions are technologically possible, at acceptable cost, but the social barriers are tremendous.

Box 11.5

Criteria for long-term climate policy as developed in the National Dialogue

Climate effectiveness: agreement exists that climate effectiveness should be the most important criterion in the stimulation of options to reduce emissions. The other criteria that came out of the dialogue may on occasion restrict the criterion of climate effectiveness.

Sustainability means social, ecological and economic sustainability (people, planet, profit). Not everybody regards the favourable options for emission reduction as equally sustainable. This applies especially for CO_2 storage, biomass and hydrogen. At the same time, it is recognized that the impacts of climate change, even if a world-wide reduction is realized in the order of 80 per cent, can be such that these controversial options cannot be neglected.

Long-term climate policy and *social support* have to reinforce each other. According to some, this means that the climate policy should follow the current developments in the sector (agriculture) and that the consumer should be actively involved in this policy (housing). A quite general impression is that important options, such as CO_2 storage, biomass and wind do not have a high score on the criterion of social support. Conversely, the options which seem to enjoy strong social support have a low score on the criterion of cost effectiveness.

Cost effectiveness: where there is a choice between options, the alternative that realizes the highest reductions for the lowest costs will be preferred. It should be noted that this criterion is mainly suitable for comparing options which are already completely developed. However, the question is to what extent this criterion can be assessed in the context of long-term developments, since the uncertainties surrounding the development of costs are enormous.

The social support for climate policy will be enhanced if options are offered which bolster the *consumers' freedom of choice*. It is assumed that the general trend will be towards more consumer sovereignty over the coming decades. This criterion can conflict with cost effectiveness, since some options will only be cost effective if they are applied on a very large scale. A hydrogen infrastructure requires large investments, but it restricts the consumers' freedom of choice.

Governmental / administrative fit refers to the preference for options which can be implemented with the current set of instruments in The Netherlands, or which fit in properly with European rules.

The criterion consistency of *governmental policy* refers mainly to the tension between climate policy and liberalization of the energy market, which was identified by different dialogue groups.

Technical reliability refers to the robustness of options. Some link this criterion to a preference for simple, low-tech options with a long life and which are easy to repair if something goes wrong. The dominant cultural bias favours high-tech, but relatively vulnerable options.

Potential for innovation means that options will be assessed in relation to their capability to generate further sustainable technological innovations. The assumption that underlies this criterion is that large-scale innovations will be needed, since the degree of sustainability of options such as CO_2 storage and biomass is doubtful.

For others, who are less optimistic about the technologies currently available, the question is whether the measures that have to be adopted in order to cut emissions up to 80 per cent, will be acceptable to society. This reveals the tension between the criterion of climate effectiveness on the one hand, and the sustainability and social acceptability criteria on the other (dilemma 3: 'climate change' versus 'sustainability'). This theme will be developed further in the next section on the basis of the options that are regarded as necessary to achieve drastic reductions, but at the same time appear to be controversial.

Controversies about crucial options for long-term reduction strategies

This section discusses the serious doubts that exist about the options that are regarded as crucial for achieving drastic reductions. In making the transition towards a climate-neutral society, policymakers will have to deal with these controversies.

CO_2 removal and storage

There is concern that carbon, once it is stored underground, may leak out at some point in time with potentially serious consequences for the local environment. Another reservation relates to the ability of government and society to manage and monitor underground CO_2 for an infinite period of time. In addition to these objections, concerns have been expressed that opting for large-scale CO_2 removal and storage will lead to a neglect of sustainable options, especially the implementation and further development of solar and wind.

Against this line of argument, it has been suggested that even if drastic emission reductions are realized by 2050, climate change may have irreversible impacts on highly valued eco-systems. In this case, CO_2 removal and storage are probably unavoidable. After all, *if the sustainability criterion warrants objections to CO_2 storage under the ground, the same criterion can certainly not lead to acceptance of an ongoing increase of CO_2 in the atmosphere.* Another possible advantage of this option is that once it proves to be cost effective, The Netherlands will no longer be so dependent on other countries and on Europe in making the transition toward a CO_2-neutral energy system.

It can be concluded from the dialogue that, under strict conditions, future CO_2 removal and storage may garner sufficient social support. However, the first condition to be fulfilled will be strong support and encouragement for those options that are expected to make CO_2 storage unnecessary in the long term.

Biomass

Biomass, too, is not generally considered a sustainable option. There is skepticism about the availability of biomass, given the volume and diversity of claims in various sectors. Concerns have been expressed that the industrialized countries, where energy use and transport are growing rapidly, will once more solve their problems at the expense of less well-to-do regions such as South America, Africa and Eastern Europe. Will food security in these regions not be endangered by a shift to large-scale biomass production? What are the likely negative social impacts of this?

Nevertheless, it is largely recognized that, at least theoretically, sufficient land is available for biomass. Starting biomass production on these (degraded) lands will bring considerable benefits to the exporting countries. Therefore, the potential problems arising from future large-scale biomass production and use will not necessarily lead to a rejection of this option. However, the concern about its social and ecological impacts dictates a plea for sustainable production and use (chain optimization) of biomass. One of the obstacles identified in this respect is the global climate regime itself, as it provides incentives for dumping biomass from the forest right into the oven instead of cascading.

Renewables

While on the one hand the dialogue anticipates a variety of problems with respect to the sustainability of clean fossil and biomass, on the other it observes that given the current state of technology The Netherlands faces limitations with respect to the potential of renewables (especially solar and wind). These limitations are partly physical in nature (too little space for wind turbines and solar panels) and partly due to the costs of back-up systems, which are expected to be huge. Nevertheless, there are no limits in an absolute sense (there is always the possibility of imports) and as systems become better linked in the future the need for back-ups is likely to decline. For a better understanding of the divergent views on the potential for sustainable options, one has to distinguish the positions of the various sectors. For the housing sector, renewables can already make a difference. For this sector, the limits to renewables have not been reached yet. No major breakthroughs seem to be required to realize the 80 per cent reduction. The options are also expected to meet with a positive response from consumers. However, for the sectors that are most responsible for current Dutch greenhouse gas emissions - transport and industry - solar and wind do not yet represent a serious alternative (now or for decades to come). Moreover, the public acceptance of renewables may also be doubtful. It is not unlikely that a hypothetical offshore windmill park with a 20,000 Megawatt potential along the coast of North Holland (from Den Helder to Zandvoort) will meet with public resistance. Is it sustainable to externalize problems from land to the sea?

Energy efficiency

Demand-side management is a crucial option for all sectors in reducing emissions. As regards the possibilities for further efficiency improvement, opinions diverge. Some are worried that the pace of efficiency improvement (about 1–2 per cent per annum since the 1970s) cannot be maintained. Existing processes are getting very close to the thermodynamic minimum. Hence, the Industry group made a plea for the strong encouragement of process innovations through broad co-operation between companies.

Breakthroughs in sustainables and demand-side management are both regarded as necessary to avoid too great a reliance on CO_2 storage and biomass in the longer term.

Besides the question of the degree to which 80 per cent reduction of emissions is feasible (given the technological options on the agenda of the dialogue groups and in a way that is acceptable to society) different insights came to the fore about the role of government in climate policy. The following section addresses this aspect, going further into dilemma 1: 'a route involving a small number of decision-makers versus a decentralized route involving multiple and diffuse decision-makers' and dilemma 5: 'Central control versus co-production'.

Government's role in achieving 80 per cent reduction

Is it possible that government can only use the market mechanisms, given the technological breakthroughs that are seen as necessary and the implications for infrastructure associated with a large number of options? Can the Dutch government and European government make a difference in an era of liberalizing energy markets? The National Dialogue observes that one impact of liberalization is that government as well as large companies show less interest in (fundamental) research. Some expect that the shift from regulatory toward market instruments, especially the introduction of a system of tradable emission permits, will in time generate the technological innovations needed. If such a system includes an emission cap which is regularly revised downwards, options that are initially too expensive can become cost effective. Buying additional emission rights might be more costly than implementing the new (expensive) technologies. Others raise doubts here,[1] arguing that government and not the market is the institution best capable of safeguarding long-term perspectives.

Government is needed particularly to stimulate and support research & development for a (limited) number of technologies and to act as 'lead customer'. Moreover, government also has a traditional role to play in realizing major infrastructure projects.

Box 11.6 summarizes different views from the National Dialogue which link insights and perceptions regarding the available knowledge and technologies with opinions on the speed needed to implement market instruments and institutions. As an ideal-typical representation, Box 11.6 fails to do justice to the dialogue, because many issues, opinions and nuances are left un-addressed. What should be specifically mentioned here is the observation from various dialogue groups that climate policies may benefit from more active involvement by consumers who are willing to make a contribution.

In fact, Box 11.6 distinguishes four problems. In cell A an innovation is available at a reasonable cost, but the problem is how to promote its adoption through the market. The solution is found in giving CO_2 a price. There are a great number of options that this problem frame can be applied to. One of these may be CO_2 removal and storage, provided that safety is no longer an issue. In cell B, no product is available yet, but there is a reasonable expectation that one will become available after a period of time. It is also clear who (which specific companies) is responsible for this. Long-term standard setting, as proposed for the development and implementation of CO_2-neutral transport fuels, is considered a market instrument that can easily evolve into a system of tradable permits. The problems identified in the right-hand column of the table can both easily be addressed with mar-

---

[1] Opinions differ on this point; see also the contributions of Arentsen et al. and Woerdman et al. in this volume.

Box 11.6

Conflicting views on the transition toward a -80 per cent energy system[2]

Emission trading regime / other market-related instruments? Technology available?	In the medium / long term	As soon as possible
Not yet	R&D through non-competitive, cross-sectoral co-operation. Major role of government in financing and R&D infrastructure. D C	Long-term standard setting for specific sectors or technologies / combine with ecotax to avoid externalities B A
Yes	To mobilize support for the adoption of innovations by non-market instruments (create lead customers).	Acceptance and implementation of options by companies and consumers.

ket instruments. The situation is different for the problems in the left-hand column of the table. The main difference is that rather than competition stimulated by market (-like) instruments, co-operation between parties is considered pivotal for solving their problems. In cell C, a product is available but for some reason (there may be a variety of reasons here, including the costs of the product or the absence of a market) it cannot be made available through market competition. Government must use non-market instruments to stimulate its adoption. The plea for a fixed share of renewables in the energy supply is an illustration of this kind of solution strategy. In cell D, the theoretical knowledge is available but a lot of R&D seems necessary to bring a product to the market. In this case – and contrary to the starting position in cell B – there may not even be a clear idea about the kind of product to be developed, the actors to be involved or the knowledge needed to achieve a breakthrough. Options that may fit in with this problem frame are breakthroughs in industrial-process efficiency, or, for the long term, solar.

So far, the problems and accompanying solution trajectories have been presented as distinct and complementary. Together, they could be regarded as sketching a trajectory for a transition over time, starting with R&D and ending with measures to endorse fair market competition. In everyday practice, however, the distinction between these kinds of problems may be far from obvious. It has already been seen that the question of whether state-of-the-art technology can help in realizing major reductions is answered differently by the sectors housing and industry. To explore the meaning of this point in more detail, efficiency is taken as an example.

---

[2] Based on M. Hisschemöller, M. Andersson, M. van de Kerkhof and W. Tuinstra: 'What we do not know yet about the institutions needed for the transition toward a decarbonized economy; A report from the COOL dialogue.' *Paper presented at the METRO Conference on Institutions and Instruments to Control Global Environmental Change*, 21-22 June 2001, Maastricht, The Netherlands: page 16.

Divergent views were articulated on efficiency improvement in the National Dialogue. On the one hand, this can be explained by the fact that efficiency is a package including a diversity of options and issues. Some are already available and can be implemented soon. Others require fundamental research and technology development. This might lead to the conclusion that the policy instruments needed to improve efficiency should therefore also be very diverse. However, other participants in the dialogue put forward opposing views, which may justify the inference that the dialogue has articulated conflicting views on one and the same issue. The view that technological barriers constitute the main obstacle at this moment can be addressed by a strategy as proposed in cell D, non-competitive co-operation and strong government support for R&D. By contrast, there is the observation that in the past efficiency has increased rapidly in response to external pressures (energy crisis, price increase). From this, one might conclude that incentives for further improvement of efficiency will be provided by a regime of tradable emission rights (cell A).

This example raises several important questions. Who decides in the decades to come what kind of problem is involved and which solution strategy fits best? Then, if it turns out that it is not a single problem but a number of different ones (in which case the views put forward are not really conflicting), how 'tailor-made' must a policy be in order to implement the best technology? How 'tailor-made' can a policy possibly be in order not to be caught up in detail? When do policy instruments conflict with each other and does policy lose its transparency and consistency? To what extent does liberalization allow for far-reaching government support if this could lead to the promotion of certain technological developments at the expense of others? The major challenge in the development of long-term climate policy probably lies in addressing these types of question at the crossroads of technology, economics and governance.

It is commonly asserted that environmental policy must always use a mix of policy instruments. A somewhat neglected issue is whether this is always possible or whether choices have to be made. The lesson to emerge from a close examination of the National Dialogue is that it is very important indeed to explore and investigate when and how different policy instruments may clash with each other.

Preparing for the long term: recommendations from the National Dialogue

This final section presents recommendations on how to prepare for the long term. These recommendations relate to actions to be taken in the period between now and 2012. They are addressed to government, business and the environmental and consumer movements. The point of departure is the general picture that emerged from the National Dialogue:

From the National Dialogue it is concluded that emission reductions up to 80 per cent by 2050 are conceivable. But not all dialogue participants are (equally) optimistic about the feasibility of such reductions. More specifically, there is considerable doubt as to whether these reductions will be possible without causing or aggravating problems other than climate change here or elsewhere. This leads to the conclusion that -80 per cent may become attainable for The Netherlands in a socially acceptable manner if, next to overcoming many social, institutional and psychological barriers, major technological breakthroughs are realized in specific areas. European and Dutch governments are supposed to take a leading role in this respect, but it is doubtful whether they can do so.

On the one hand, this conclusion implies a mild optimism, but on the other persistent doubts and concerns with respect to the desirability of certain options and the capabilities of government. An effective long-term government strategy must find a way to deal with these contradictory impressions, with the evident doubts and diverging opinions. After all, they all point to real issues that must be coherently addressed. Recommendations 4–10 below deal with these issues. The dialogue also generated some shared views as regards general factors that will make a long-term climate policy successful. These are dealt with in recommendations 1–3. Box 11.7 provides additional information about specific recommendations for the different sectors developed by the dialogue groups.

General

1. *Use the coming 10 years well!* It is already possible to start implementing various options for the four sectors. But although the climate problem is urgent and its urgency is likely to increase in the coming period, it is particularly important to take sufficient time to work out a coherent long-term strategy and take action accordingly. In this respect, it is critical to fine-tune substance and process, and to co-ordinate actions at the global, European and national level. The timing of decisions (not too late, but not too early either) is also of critical importance.

2. *Enhance a sustained involvement with the issue!* It is essential to create the necessary conditions for the transition toward a CO_2-neutral energy system. These conditions include:
    - a sense of urgency among stakeholders and the public at large over a long period of time about the need to tackle this issue,
    - a sustained political will to make sincere efforts (high on the political agenda for decades),
    - consistency of government policy, providing private parties with the encouragement to make their contribution.
    Long-term policy must be developed in addition to, but not in isolation from, existing short-term policy. Long-term policy must allow for critical evaluation of, and reflection on, short-term actions. The approach taken in the COOL project, a dialogue between science and stakeholders from society on the long term, may serve quite well to do this. It is therefore important to learn from the lessons of the COOL process.

3. *Social support for -80 per cent as a long-term target?* The National Dialogue addressed the feasibility of emission reductions up to 80 per cent, not their desirability. The findings of the dialogue and the involvement of its participants justify asking the question of whether to establish a long-term reduction target in the order of -80 per cent in a national as well as a European context. Sufficient time is needed to

fully explore the implications and impacts of such a decision. Answering such a question supposes that the specific themes below are coherently addressed.

Themes for long-term climate policy

4. *The effectiveness of instruments and institutions over time:* The National Dialogue signals a relation between the technological, economic and political / institutional aspects of long-term climate policy. It is critical to gather information about the effectiveness of instruments and institutions needed to realize reductions in the various sectors. At the moment, there is no integrated, multidisciplinary approach to this issue. An integrated assessment, taking place at the crossroads of technology, economics and governance, could shed new light on the interactions between the choice of instruments and (unwanted) policy effects over time.

5. *Knowledge, knowledge infrastructure:* The dialogue suggests that it is crucial to invest heavily in fundamental and applied research to force major breakthroughs in a European context. One suggestion is that governments, business and science collaborate to establish an international Institute of Excellence which would work on optimizing solar PV. Similar initiatives may be taken with respect to hydrogen and biomass. These proposals from the dialogue are at odds with the tendency of the Dutch government and business to withdraw from research at this moment. Efforts to stimulate R&D must be accompanied by a critical reflection on the current knowledge infrastructure, especially on its transparency and its accessibility for parties who want to conduct or use research.

6. *Market development and the adoption of innovations:* In addition to proposals for a strong impetus for R&D, the dialogue indicated barriers that prevent innovations from penetrating the market. Government traditionally assumes the role of lead customer; that is, it takes the risks linked to novel products. Besides government, environmental and consumer NGOs may also be of significance here, e.g. by mobilizing lead customers who may then become shareholders in the new product.

7. *Implications for space and infrastructure:* Closely related to recommendations 4 and 5, it is critical to get a clear picture of the implications of climate options for the use of scarce space and infrastructure. Such an assessment would have to focus on a range of issues, such as expansion of the electricity grid, underground or surface grid and pipelines, requirements for gas stations, and numerous other implications with respect to waterways, bridges, landscape architecture, design and management of industrial areas, opportunities for (re)locating businesses and the like. There is still much to be learned about costs, sustainability, as well as the division of labour between government and private parties in adjusting infrastructure to climate policies.

8. *Empowering consumers:* The potential of consumers as actors in the climate issue has so far been underestimated in Dutch climate policy. Attention for consumers largely focuses on providing incentives for individuals to change their behaviour.

But (organized) intervention by consumers could provide a stimulus to accelerate changes on the supply side. Existing opportunities to increase consumers' involvement should be explored. Strengthening individual responsibility and consumer sovereignty must be pivotal objectives in such an assessment. One instrument that needs to be further investigated is providing individual consumers with emission rights.

9. *Trajectory Biomass:* In this trajectory, all barriers to large-scale adoption of biomass are looked at in an integrated way. At the international level, initiatives are needed to develop a system for certifying sustainably-produced biomass. One of the major topics should be whether chain optimization is a workable national and international objective (at this stage the climate regime provides incentives for burning biomass). This question also implies opportunities for developing an international monitoring system. Barriers for bio fuels (such as fees for alcohol), must be inventoried and addressed nationally and at the European level. Research into various biomass potentials must be intensified.

Box 11.7

A selection of specific recommendations for the four sectors

Housing
- Integrated design, include the use of passive solar energy in building codes;
- Support solar PV by stimulating agreements between solar PV producers and energy suppliers about liability and incentives (government);
- Stimulate the use of wood in house building;
- Address the issue of groundwater contamination by antifreeze in heat pumps, generic subsidies (government).

Industry and Energy
- Measures to stimulate CHP generators in industry, trajectory for linking CHP generators with clean fossil options (government, companies);
- Develop plans to implement a hydrogen infrastructure (step-by-step introduction or radical change), (all parties involved);
- Develop industrial parks in accordance with ecological principles and create agreements between companies to coordinate the demand and supply of heating and cooling (sector).

Agriculture
- Gradually diminish production and use of artificial fertilizer (Netherlands, sector, EU);
- Co-ordinate policies regarding nature and landscape with those regarding the climate issue (sinks) (Netherlands, managers of nature reserves);
- Create good house-keeping on the farm (sector);
- Make an inventory of sustainable food options (all parties involved).

Transport
- Develop long-term standard setting for the transition to a CO_2-neutral fuel market (at European level);
- Internalize the costs of CO_2 in the fuel price (at European level);
- Introduce a test for the climate impacts of new infrastructure (Netherlands);
- Include transport as a part of environmental management systems and certification schemes (sector).

---

10. *Trajectory CO_2 removal and storage*: This trajectory delivers all relevant information for political decisions on CO_2 removal and storage. Special attention must be given to safety issues related to transport and underground storage, as well as to possible reactions underground. Demonstration projects may be started. The assessment and decision-making should specify under what strict conditions (e.g. safety, continuation in investments in renewables), when and at what kinds of locations CO_2 storage will be permitted. Given the nature of the resistance to this option and at the same time the opportunities this option may provide to significantly reduce emissions at reasonable costs, for this trajectory it is even more advisable to have an open dialogue that is able to define alternatives for decisions.

Box 11.8

Engaging stakeholders: process lessons from the National Dialogue

Except for the content-related outcomes of the National Dialogue, it is interesting to mention some process-related findings. Exercises like the National Dialogue can be a helpful vehicle in exploring the different steps in the transition to a climate-neutral society. Therefore, it is important to learn from the process lessons of the National Dialogue.

- A dialogue group with stakeholders from different fields of expertise, with different opinions and views, will increase the possibility to generate new insights for policy.
- The success of the group depends heavily on the quality of process support. It is critical that the different steps in the process are transparent to the participants.
- In a demand-driven dialogue, it is crucial to choose as a starting point for the discussion the wishes, concerns and expectations of all participants.
- A certain degree of autonomy of the dialogue groups will increase the participants' involvement in the work of the group and in the end product.
- The role of scientific support in a dialogue deserves special attention. The information that is offered should be accessible, compact and tailor-made. Furthermore, a proper communication of information is of vital importance.

- The backcasting methodology results in insights related to opportunities for and obstacles to the implementation of options for climate policy. However, it does not by itself provoke the participants to articulate and discuss conflicting views.
- Using future images and backcasting stimulates a long-term scope in the dialogue. These methods do not cut off the participants from their own experiences, opinions and interests. This should not be the project's intention.
- An extended preparation phase and a good budget are crucial for a proper organization and management of the dialogue.
- To work in an interdisciplinary team requires careful communication and a good working plan. This takes time!

References

Faaij, A., S. Bos, J. Spakman, D.J. Treffers, C. Battjes, R. Folkert, E. Drissen, C. Hendriks and J. Oude Lohuis, Beelden van de Toekomst. Twee visies op de Nederlandse energievoorziening ten behoeve van de Nationale Dialoog, 1999.

Hisschemöller, M., The National Dialogue (COOL). Results and recommendations. Synthesis Report, IVM Amsterdam, nr. O-01/12uk, 2001.

Hisschemöller, M. and M. van de Kerkhof (eds.), Climate OptiOns for the Long term – Nationale Dialoog. Deel B – Eindrapport, IVM Amsterdam, nr. E-01/05, 2001.

Hisschemöller, M., M. Andersson, M. van de Kerkhof and W. Tuinstra, What we do not know yet about the institutions needed for the transition toward a decarbonized economy; a report from the COOL dialogue, paper presented at the METRO conference on Institutions and Instruments to Control Global Environmental Change, 21-22 June 2001, Maastricht, The Netherlands, 2001.

Chapter 12

The climate-neutral society: opportunities for change

Walter Vermeulen, André Faaij, David de Jager and Marcel Kok

Introduction

If there is one thing that has become clear from the contents of this book, it is the complexity of the transition processes that are inevitably linked to realizing a society that is (almost) climate-neutral. Whatever the strategy to reach the substantial reduction of greenhouse gas emissions by 80 per cent in 2050 compared to 1990 levels, the implications are imposing. The overall objective of this book is to explore the changes in social trends needed for long-term substantive reductions in greenhouse gas emissions and to assess the prospects for realizing such trend breaks in society.

Numerous technological or macro-economic studies have been performed, identifying both options and strategies that contribute to a climate-neutral society. In its role of conducting research for policy, the Dutch National Research Programme on Global Air Pollution and Climate Change (NRP) has supported a number of projects examining the elements of a 'climate-neutral society', thus exploring routes of greening society. The focus in these projects is often on analysing new technologies capable of enabling a drastic reduction in greenhouse gas emissions. In essence, these future studies consist of expert opinions about new technologies (the tempo of innovation and diffusion and the feasibility of their application). As such, these research activities combine technological, environmental and (in many cases) economic knowledge in forecasting models. Economics comes into the picture if learning curves, describing the reductions in costs of innovations over time, are included in the models. The effects of policy mixes are also assessed in some of these studies. They tend to present our salvation through technology, with a touch of economic instruments or regulations added, depending on the scientific discipline of the authors.

Within the NRP, a range of social scientists have also been working on social causes, responses and possible solutions. This has provided an opportunity to comment on these techno-economic images of the future from the perspective of required social change. We invited these researchers to join us in a *'back to the present'* exercise: commenting on what would need to change in order to arrive at these alternative futures. Reasoning backwards (or 'back-casting') in this way is relatively easy, enabling one to identify institutional and cultural barriers and the like. Of course, in doing so, we had to look into an uncertain future – into an ever-dynamic and changing social, political, international, economic and technological environment. Recognizing and accepting this societal complexity may very well be the first crucial step towards managing the major changes required if the objective of protecting the climate over the long term is to be realized. This book shows that, in terms of technological options and further development and implementation of policies that can contribute to the key objective, there are countless possibilities, opportunities and starting points throughout society. We have learned much about existing barriers in the domains that have been discussed and analysed, which in turn provides some insight into how to overcome them. We have also seen promising developments in changing practices. In this final chapter, we will be looking forward again, discussing the findings of these confrontations and trend breaks and evaluating the most important dilemmas. We focus on some critical issues and opportunities for change, being aware that our final remarks may offer more food for thought and further research than providing answers to questions.

Social trend breaks: opportunities

Looking back at the analysis in the previous chapters, we can see that required changes in social trends have been identified, as well as opportunities and instruments for effecting these changes. The 'back to the present' exercise highlights the contrasts between the social implications of the future visions and the situations that exist today. In this chapter, we will look at some of the main conclusions that affect the spheres of consumption, production and, consequently, governance. We will discuss just a few of the issues covered in the book – the subjects chosen here serving merely to illustrate the line of reasoning that arose from the 'back to the present' exercise.

Consumption

Looking at the sphere of consumption, visions A and B in Chapter 2 are constructed on the basis of their performance in delivering sufficient reductions in greenhouse gas emissions. It is expected that there will be changes in *household consumption* in reaction to transitions in the areas of product development, food production, housing and transportation. In the future scenarios, such changes are simply presupposed, not analysed. Recent environmental policy has addressed citizens as consumers, but with only marginal results. Chapter 5 showed that in practice, the variation in the total environmental impact of the consumption packages of individual consumers can be quite strong, even within income categories. Moll

---

Box 12.1

Required trend breaks, opportunities and instruments in the area of consumption

Required social trend breaks	Addressing various segments differently (Ch. 3, 5)
Opportunities	Willingness to act for part of population (Ch. 5)
Instruments	Direct routes: information, pricing, standards
	but: diversification of approaches (Ch. 5)
	Indirect routes: producers, physical infrastructure (Ch. 5)

and Groot-Marcus in Chapter 5 discussed some experiences with the perspective project and ecoteams that could help consumers to assess the environmental impact of their consumption and redirect their routines. This can be effective in the short term, but research shows that consumer appreciation of consumption alternatives does not correspond with their environmental impact. As discussed by Moll and Groot-Marcus in Chapter 5, some of the behavioural alternatives with the greatest impact are perceived by the public as indispensable, whereas some with small impact are perceived as easily replaceable (as shown in the study of Uitdenbogerd et al.).

Providing consumers with information to allow them to assess environmental impact themselves has a limited effect. The public response to such information is interpreted by consumers within the context of their own value system and worldview, which are, by nature, very diverse. The argument here is that using strategies based on awareness and rational decision making is only effective for some of these cultural-normative groups.

What does this teach us about the way to achieve a climate-neutral society? It opens two routes for reducing the ecological impact of consumer behaviour. The first avoids dependency on consumer (and voter) behaviour. Instead, it depends completely on producers improving their products, reducing their environmental performance and ensuring full market breakthroughs by means of financial instruments. In this case, policies for climate change will have hardly any effect on the everyday life of citizens. The second route depends on the social responsibility and moderation of informed consumers. In this approach, the government addresses consumers, either directly or indirectly, by means of information about the necessity of changing consumption patterns and with product-related information (such as ecolabels, information services, stimulation supply, etc.). The risk of the first route is that it can result in reduced public commitment, market pressure and political attention to issues on climate change. The second route is risky because it does not consider the variety of consumer attitudes, behaviour and values.

In regard to consumer commitment, an interesting development occurred during the writing of this book. The introduction of a green electricity scheme in a liberalized energy market in The Netherlands in the summer of 2001 resulted in a growth of the market share of green electricity from 4 per cent to 12 per cent of all Dutch households in only five months! This development can be examined from various perspectives (see Box 12.2), but it shows one thing very clearly: intentionally changing institutions and the roles of social actors can create change, even in consumer behaviour.

Yet, we have to keep in mind the importance of existing values and worldviews, which act as a substrate on which perceptions of environmental issues, attitudes and factual behaviour flourish. In many cases, these values and worldviews may explain why issues on climate change are disregarded or rejected. If these different values and worldviews are not taken into account, the result can be only marginally effective policies. But we must also note that in the long run, values and worldviews can be challenged. They do, in fact, evolve along with new technological systems and their accompanying institutions. Knowing this opens the way for the deliberate design of new technologies and products along with efforts to act upon their social interpretation and appreciation. And more: the variety of values

Box 12.2

Going Dutch for green electricity

'Green energy' is a general term for renewable energy that private persons and companies buy from their energy supplier. At the moment, green energy refers to electricity from renewable energy sources. However, in the future it may be possible to buy 'green gas' or 'green heat'. Those who want to use renewable energy could decide to place a wind turbine or solar system on their roof. However, not everybody has the time, knowledge, skills or space required. And it is often more advantageous to produce renewable energy on a large scale or at more optimal places. Therefore, some years ago, energy companies, encouraged by environmental NGOs, started offering renewable electricity to their customers, as an alternative to electricity generated with coal, gas or oil.

The market for green energy is growing rapidly. In The Netherlands, households and companies bought 350 million kWh of green electricity in 1999. In October 2001 this figure had grown to 2.4 billion kWh (still only about 2 per cent of total electricity consumption), with companies purchasing over 200 million kWh and about 700,000 households (about 10 per cent of Dutch households) participating. In May 2001 the number of households was only 250,000, less than 4 per cent. So, in five months, the number of participating households almost tripled. In two years, the green energy market increased by almost a factor of seven!

This enormous increase is a consequence of the tax benefits that are given to both producers and (notably small) consumers of green energy and the fact that this 'green' part of the energy market was being liberalized as of July 2001. Because of this, some retailers offer green energy at the same price or even below the price of 'grey" electricity. Combined with support from environmental organizations (e.g., the World Wide

Fund for Nature and Greenpeace) and increased marketing efforts by energy distribution companies to assure market shares in the liberalized market, the number of green clients has boomed.

The green energy market is facilitated by a system of green certificates, which enables the separate trade of electricity and certifies the 'greenness' of this electricity. It also ensures that each certificate that is being sold to final consumers was actually generated by a renewable energy source. Since the Dutch scheme allows for import of certificates (under certain conditions), it, in combination with the renewable energy certificate system (RECS) at the European level, is in fact an international certificate trading system that could be transformed into an emission (permit) trading system.

The opening of the market for green energy to small consumers (resulting in competition between retailers), the financial support on the demand side (resulting in interesting margins for the retailers), and the creation of a level playing field for energy retailers by the introduction of a certificate system, were the main drivers of this rapid trend break in the green energy market.

However, one can still criticize elements of the system: the effectiveness of the system is impressive, but it is to a large extent the effect of the tax benefits (notably at the demand side), equivalent to more than € 150 per ton of CO_2. This is relatively expensive and inefficient compared to other instruments supporting greenhouse gas or energy saving options, and it is questionable whether the problems on the supply side will be solved in the politically desired timeframe (because of problems related to NIMBY, licences, etc.). Furthermore, the system is not designed to support innovation as such (only if the market asks for it, e.g., solar photo-

Box 12.2 continued

voltaics is sometimes included in the green energy portfolio), which could result in a technological lock-in.

On the other hand, the tax benefits are financed with revenues from the Dutch regulating energy tax on non-renewable energy.

Such combined taxation and tax exemption schemes have proven to be effective in other fields of environmental policy as well (e.g., introduction of lead-free gasoline and catalyst cars).

Source: from www.greenprices.com (November 2001).

and world views within segments of society can and should be used as the starting point for designing a variety of strategies.

Such an approach does not allow us to amend the mixture of technologies in both visions of the future directly, but it does involve a trend break in the routine sequence of designing policies: starting from a techno-physical and economic approach and resulting in a policy mix of technical measures, subsequently provided with instruments for implementation. Following this argument, social, cultural and institutional criteria should be included in the assessment of mixtures of technical measures.

Production

Visions A and B in Chapter 2 depend largely on technological transitions within the sphere of production, including energy production and distribution. Looking at industry today, changes such as clean process technologies, clean fossil fuels (in combination with CO_2 storage), improving energy efficiency, ecodesign and improving material efficiency and organizing take-back schemes are required. This type of change is not entirely new. As a matter of fact, the 1990s showed a trend towards the greening of production pursued as a co-responsibility of market and state. In this development, both the modes of governance and the perception and methodology of improvement of environmental performance have changed drastically. In the mode of governance, traditional central governance by means of coercion and incentives has been modernized and extended with two complementary strategies: the first being a strategy of interactive management and internalization and the second being a strategy of enabling self-management (Vermeulen, 2002).

However, in order to reach the overall level of an 80 per cent reduction in emissions, there needs to be an enormous increase in effort. Industries going in this direction will have to deal with major challenges: it will require increased collaboration among producers along the entire production cycle, going back all the way to raw material extraction abroad. This implies co-operation among mining firms, firms supplying semi-finished products, final producers, traders and recycling firms. A second indispensable step will be the promotion of system innovations (see Chapters 4 and 7). Experience in these fields is growing but tends to be restricted to either one step back in the production cycle (working together in product partnerships) or one step ahead (i.e., organizing the collection and recycling of products).

Box 12.3
Required trend breaks, opportunities and instruments in the area of production

Required social trend breaks	Increased attention for dematerialization through life-cycle management (Ch. 7)
	Dissemination of climate-oriented designs for environment (DfE) and system innovation (Ch. 7)
	Increased market demand as stimulus (Ch. 7)
	Increased attention to threats of ICT energy savings (Ch. 8):
	• rebound
	• suboptimal organization and infrastructure
	• insufficient information to user
	• lack of co-operation
Opportunities	Positive effects sufficiently demonstrated (Ch. 7)
	Eco-efficiency of system innovations (Ch. 7)
	Easy integration in existing policies (Ch. 7)
	ICT can make minor contribution to preventing climate change, but can facilitate other processes (Ch. 8)
Instruments	Standards for recycled material (Ch. 7)
	Take-back systems (Ch. 7)
	Sustainable consumption policies can create market demand (Ch. 5, 7, 11)
	DfE practices in ICT innovation practices (Ch. 8)

Such forms of co-operation are perceived by innovative firms as opportunities, but they are far beyond the (perceived) span of control of the majority of firms that are using defensive innovation strategies.

For these firms, activities aimed at co-operation over a product's life cycle and product stewardship bring along new uncertainties with unclear commercial benefits. Discourses on the political level and public debate on policies aimed at climate change do not create an incentive for these groups of firms to get involved in such activities. For them, getting on the road to the greening of production is clearly a social dilemma. With enough reasons not to start (such as unclear economic gains, lack of experience and uncertain market demand), why be the first to do so?

From their perspective, there is no clear, collectively supported, long-term target in the field of climate policy, which strengthens this attitude among the majority of firms to wait. But this inertia can be broken down. Experience with the interactive management policy strategy in The Netherlands in the 1990s shows that breaking down such inertia in the absence of long-term perspectives is possible and may offer a way out of the social dilemma (Vermeulen and Weterings, 1997; Glasbergen, 1998; Driessen and Glasbergen, 2002). In The Netherlands experience in the 1990s, in order to establish a consensus on long-term environmental targets for specific industrial sectors and ways to achieve them, the initial step towards industry was essential. This included organizing mechanisms of institutional learning, reporting and monitoring to ensure feedback on the course of the process. The experience with the COOL dialogue (Chapter 11) shows that for certain sectors, discussing long-term climate targets and their implications can be a fruitful part of a long-term strategy.

Co-operatively defining long-term targets and ways to achieve them may be the essential first step to bring the majority of manufacturers in the production sector together to collaborate over the product life cycle, ecodesign, system innovations, material efficiency, etc. Some research on 'chains of producers, sellers and users' of consumption goods is currently underway (Ministerie van VROM, 2000, p19). Systematically addressing all relevant groups of products by collaboratively defining sustainable production for each 'sector' would be trend breaking.

Looking for the outcomes of the future visions for the agenda of social environmental science emphasizes the need for research in understanding, explaining, designing and managing processes of long-term innovation (focusing on its social and institutional context). This has recently been addressed as 'transition management', which, as a form of facilitating governance, is discussed in Chapter 4.

Looking back at the future visions in Chapter 2, transition management calls for abstaining from using future scenarios to identify allegedly 'optimal mixes' of technical measures on the basis of techno-physical and economic assessments. Transition processes will benefit from pluriformity in learning and competition between various trajectories.

Governance

The discussion above brings forward the question of the implications for governance. The visions themselves do not contain governance strategies, the relation between state and market or the role of civil society. In the context of the 'back to the present' exercise, various authors reflected upon the role of general economic instruments, the implications of the legal system and the role of local government. The reflections on developments in the field of producers and consumers, given above, also have implications for governance.

As a general line arising from these reflections, we can say that changing towards a climate-neutral society requires the organization of institutional learning processes, offering guidance and creating a common ground by means of long-term targets. This is in contrast to advocating forms of central management on the basis of coercion and economic incentives. On the other hand, the aforementioned plea for consensual governance might be unbalanced if it implies abandoning all forms of general regulative and financial instruments.

One of the discussions in this area is the role of pricing greenhouse gas emissions (see Chapters 5, 7 and 11), particularly in regard to permit trading in realizing a climate-neutral society. Energy pricing is an important condition for successful climate-change policy, but low prices make it an uphill battle. Some scholars in the field tend to be very optimistic about the ability of permit-trading schemes to ensure the reduction of greenhouse gas emissions. This is based on the inclusion of a cap (an absolute maximum) on allowed annual emissions, decreased yearly. In Chapter 9, Woerdman et al. have designed a feasible permit-trading scheme. They argue that the market will ensure that emissions are reduced in the most cost-effective way. Their argument goes beyond that, claiming that the

Box 12.4
Required trend breaks, opportunities and instruments in the area of governance

Required social trend breaks	Technology policies address system innovations (Ch. 7)
	Linking short-term policies with long-term policies (Ch. 7)
	Introducing new policy paradigms, based on ET (Ch. 9)
	Changing towards a legal system based on the social legal system approach (Ch. 10):
	▪ performing 'legal consequences analysis' in climate policy studies (Ch. 10)
	▪ giving a prominent role to local authorities (Ch. 6)
	▪ ensuring local political attention (Ch. 6)
	▪ improving local external integration (Ch. 6)
Opportunities	Eco-innovators as drivers of industrial transformation processes (Ch. 7, 11)
	Administratively efficient ET scheme (Ch. 9)
	Policy attention in 40 per cent of municipalities (Ch. 6)
	Potential effectiveness of local policies (Ch. 6)
Instruments	Instruments aiming at (Ch. 4):
	▪ technology variation
	▪ actor interaction and learning
	▪ market selection
	▪ strategic niche management
	Emission trading for companies, households and transport (Ch. 9)
	More government responsibility (Ch. 10):
	▪ norms and instruments of public law
	▪ strong enforcement
	▪ strong public participation
	▪ broad legal protection
	Organizational and process improvements (Ch. 6)
	Climate policy management systems (Ch. 6)
	Participatory policy-making tools available (Ch. 6, 11)
	Pricing (Ch. 5, 7, 11)

complex policy mix of central management, interactive management and stimulation of self-management should be abandoned – thus pleading for a major administrative trend break.

Such debates challenge the emphasis on consensual governance. Can a sweeping statement like the one made above be substantiated? Can a central management strategy be promoted solely by incentives, relying on the invisible hand of the permit-trading market? This idea has found its way into the implementation of the Kyoto agreement, where emission-trading markets are being organized.

One major obstacle to permit trading, in the stubborn practice of political decision making, would be the essential feature of annual decreases in the absolute maximum of allowed emissions. But even supposing that this can be done, it remains to be seen whether price incentives alone will be enough to speed up the dissemination of climate-neutral appliances and renewable energy in industries and households. Looking at research on the diffusion of energy and environmental innovations, economic evaluation (prices) is only one of the variables explaining the pace of diffusion. Placing these issues on the business agenda, differences in business innovation strategies and innovation capacities, along with features of the adopter's socio-institutional environment, are as relevant to the rate of diffusion as pricing is. One example is that the legal system of environmental legislation is mainly based on stationary norms and protects the rights of industry once regulations and permits are issued. This hampers the opportunities for governments to impose more stringent rules on a regular basis. The discussion about the difficulties of making the step from incremental innovations to system-wide innovations should also be kept in mind here. Although they can provide a strong incentive, price incentives alone are not enough to elicit such changes in everyday routines.

Criticizing an exaggerated belief in economic instruments should, however, not distract our attention from another important issue: what would be a proper balance between central steering (including pricing of CO_2 emissions) and collaborative strategies for steering? Social, policy and economic scientists often tend to take a position at one side of the dichotomy or the other, stressing the imperfections of the opposite side. We should recognize that in practice both strategies are applied, affecting the individual behaviour of organizations and individuals either directly or indirectly. Research on assessing these effects comprehensively, focusing on the mutual effects of both strategies, deserves more attention (the importance of existing rules and incentives as a substrate for consensual strategies and vice versa: the importance of consensual strategies as a substrate for greater acceptance of, and compliance with, rules and the recognition of financial incentives).

Future scenarios, as discussed in Chapter 2, do not include discussions about the effects of the strategies covered in this section. But by their very nature, they elicit reflections of the kind shown in this book. Thus, such a 'back to the present' exercise forms an invitation to develop this kind of social science reflection in a more systemic way.

Dealing with social dilemmas

The discussions above show us that technological answers, as suggested in the future visions, are a first step in exploring strategies for a climate-neutral society. They serve to inspire imaginable futures. Often the step after analysing various alternative futures is to select the most optimal mix. In this book, the opposite is done: on the basis of environmental social science reflections on the future visions, arguments to develop as much variety in technological options as possible and to create pluriformity are made. It is clear that an analysis of the social consequences of technological visions of a climate-neutral future leads to a response that is different from the seven dilemmas put forward in Chapter 3, and different from responses that might arise from a techno-economic evaluation of such scenarios.

Maybe this is most evident for dilemma 1: a *'small number of decision makers' versus 'decentralized, multiple and diffuse decisions'*. It might seem to be attractive to follow a line of reasoning oriented towards technological feasibility, environmental effectiveness and cost-effectiveness focusing on large-scale solutions within the control of national governments. However, the visions presented in Chapter 2 suggest, in themselves, that we will need a large variety of options. Focusing on a small number of large-scale technologies may be attractive in the short run, but in the long run, an 80 per cent reduction in emissions will have to come from a wide array of options. Moreover, it is unlikely that other actors in society will not be needed to make their contribution in their roles as producers, consumers, investors, decision makers, motorists, voters, etc. For example, in the COOL project, it was concluded that the role of consumers in climate policies should be strengthened. Consumer demand may be able to accelerate technological change and to change supply. Moll and Groot-Marcus (in Chapter 5, on households) tend to be somewhat pessimistic about the willingness of consumers to take on this role, but at the same time, they also see various opportunities for (indirect) policies to realise climate-neutral consumption patterns in households. Consumers, in their role as citizens, are also crucial to sustaining a political willingness to deal with the climate problem, even if social costs for (at least) some actors will increase.

Stressing multiple trajectories brings us to the question of governance addressed in dilemma 5 as the *contrast between central steering and co-production*. In our view, the issue is to find a proper balance. Arentsen et al. (Chapter 4) ask whether a generic approach on its own can promote the development of radical solutions. Many of the authors advocate an approach that combines co-production and co-evolution to get beyond incremental improvements and to achieve system innovations to deal with the climate problem. Another reason for tailor-made approaches is the need to address rebound effects and the effects of growing wealth. As pointed out in the chapters by Moll and Groot-Marcus (Chapter 5), Hekkert et al. (Chapter 6) and Slob and van Lieshout (Chapter 7), economic growth often counteracts the environmental effects of options and policies, resulting in an absolute increase of emissions. Long-term policies may also need long-term standards and proper monitoring systems to deal with this problem (as suggested in the COOL dialogue to introduce a long-term standard for CO_2-neutral transport fuels, for instance).

In contrast to the plea for co-production and co-evolution is the idea, proposed by Woerdman et al. (Chapter 9), that emission-permit trading will be the only policy instrument necessary to bring about an 80 per cent reduction in emissions. As argued in the previous section, putting our trust solely in central steering and market incentives will not be effective. This debate also came up in the COOL dialogue (Chapter 11), where a tailor-made approach, matching the stages of technology development, was suggested.

With respect to a differentiated, tailor-made approach aimed at co-operation, the 'persistence' of a differentiated approach over a long period of time (demanding sustained political and societal willingness to deal with the problem) and the effectiveness of such policy strategies to deal with a large number of small actors can be questioned. However, this argument is also relevant for emission-trading schemes, relying on a sustained political and societal willingness to maintain periodic reductions in the national emission cap. Here, a proper balance of strategies may bring mutual advantages: co-production enabling learning in specific social sectors, thus contributing to a higher level of commitment, and central strategies with market incentives providing economic motives and a mechanism for achieving eco-efficiency.

This notion of balancing central steering and co-production may also offer guidance with respect to dilemma 3: *'focusing on climate change' versus 'pursuing sustainability'*. It may be necessary to focus on climate change to ensure the attention and commitment of the various implementing actors in society. On the other hand, from the perspective of businesses, consumers, local authorities, etc., this does not seem wise. Consumers do not perceive climate as a separate issue (Chapter 5). Businesses work with integrated assessment methods and management systems (Chapter 7). Local authorities apply, or should apply, integrated environmental policies when 'enhancing external integration'. These tendencies require policies aimed at sustainability in the broad sense instead of focusing on climate change as a separate issue. But this refers mainly to such things as operational policy instruments and assessment methods. Here again, we suggest a proper balance. Where we advocate tailor-made approaches and dialogues on climate policies with specific sectors, co-operation can very well focus on climate change, relevant options for reducing greenhouse gas emissions and their implementation strategies. When such dialogues have led to shared visions on the application of options, instruments for implementation will be required. These would then best be linked to existing integrated policy instruments, assessment methods and management systems aimed at sustainable development.

The fourth dilemma of *'demystification' versus 'need for increased awareness of change impact'* has not been dealt with explicitly in this book. One disadvantage of generic, mostly 'invisible' solutions and instruments, with a focus on large-scale technological trajectories, may be that climate change will not get the level of political and public attention required for sufficient diffusion of other necessary technologies. Chapters 5 and 6 both show acceptance of and motivation for change within groups of consumers and municipalities. But the other side of this coin is that far larger proportions of these consumers, of local politicians, etc., are not very interested. This calls for increased awareness, but not in the form of apocalyptic messages that will not fit their frames of reference.

A solution here may be contained in dilemma 6: *'transition through co-benefits' versus 'inevitability of a special approach to climate change'*. Many of the chapters have touched on this issue. It always seems wise to start promoting options that generate co-benefits. Stressing co-benefits and giving priority to options with clear co-benefits may be helpful in persuading groups that would otherwise not be successfully approached into adopting these innovations. Yet, the discussions in this book do not provide any clues to whether options with clear co-benefits together have sufficient reduction potential. In the literature, estimates of the magnitude of the net ancillary benefits and costs of greenhouse gas emission reductions in industialised countries vary widely. But according to the Organisation for Economic Co-operation and Development, even the most conservative estimates suggest that they are significant and may offset as much as one-third of the abatement costs for modest mitigation efforts (OECD, 2001).

No direct reference has been made to dilemma 2: *'acting now' versus 'delayed response'*. Arguments for delayed response mainly rest on technological calculations and expectations about the speed of diffusion. If one thing has become clear in this book, it is that there are many socio-cultural obstacles to the dissemination of technologies. This calls for maximum variation and competition between technological trajectories, instead of relying on an optimistic belief in technological progress alone.

The final dilemma, number 7 *'decentralization and Europeanization' versus 'the need for national direction of the transition'*, raises the question to what extent national governments in Europe (especially of small and medium-sized countries) will have enough independence for decisive national policy making. As discussed in Chapters 4 and 6, governments at the national and European level are supposed to play a leading role in stimulating innovation and facilitating social transitions. On the other hand, some of the proposed generic instruments, such as permit trading, require implementation at only the European level. It doesn't make much sense to set a Dutch national standard for CO_2-neutral fuels when it is clear that on this level, a European approach is needed. System-wide changes would seem to require even more international co-operation than an incremental approach. Relatively small countries, but front-runners like The Netherlands and the Scandinavian countries, may take a leading role here by showing success from innovative approaches.

These examples illustrate the importance of finding the right level of action. But as the chapters on households (Chapter 5) and local authorities (Chapter 6) show, there is also a great deal of room for policies at the local level, as well as the national (the chain-management approach, advocated by Hekkert et al. (Chapter 7) and technology policy strategy advocated by Arentsen et al. (Chapter 4)). The COOL approach, as discussed in Chapter 11, in which a combined local and national approach is advocated by some sectors, could provide long-term direction for local authorities.

It is clear that with these general approaches to climate change, the *'back to the present'* exercise results not in new future visions, but rather, in amended visions made up of mixes of technological infrastructures. The virtue of the exercise is in exploring social constraints and thus setting new agendas for environmental social science and for policy-making.

Key issues in the transition towards a climate-neutral society

At this stage, finalizing our *'back to the future'* exercise by looking forward again, we want to discuss three issues that, in our view, are the main issues coming out of our reflections and discussions during the process of writing and editing this book:

- organizing the transition towards climate-neutrality as long-term learning and feedback
- balancing generic central steering and specific co-production
- handling conflict and equity.

Long-term learning and feedback

In our efforts to move in the direction of a climate-neutral society, it is essential to adopt a long-term perspective and to enable a collective learning process in society. The visions that were presented in Chapter 2 of this book are a minor example of such long-term perspectives. Starting from clear, long-term targets for reduction at the sectoral level and translating them into development trajectories over time (e.g., by applying back-casting techniques), the consequences of short-term actions over the long term become very apparent. The debate should focus not on choosing selective packages of technologies, but on enabling pluriformity and competition.

In an environment that constantly changes over time, while long-term goals remain fixed, it is necessary to regularly update and check the level of success of short-term policies. When seen in this way, it is not a matter of fixing a strategy for the coming half century, but rather a matter of agreeing on procedures to keep track of long-term objectives over time, in the face of the inevitable unexpected developments.

The dynamics of managing long-term transition and, at the same time, maintaining long-term goals, requires intensive and almost continuous dialogue with relevant and diverse groups of stakeholders. Many chapters in this book and, in particular, the experience gained in the COOL dialogue described in Chapter 11, indicate the need for continuous communication about interests, perspectives, possibilities and responsibilities among a multitude of actors. An ongoing dialogue (as a learning process) among the actors involved may help:

- further develop a long-term vision and social strategy for achieving a climate-neutral society
- enable stricter targets over a long period of time
- enable governments to develop consistent policies and learn which options should be favoured for the long term in order to realize institutional change and to anchor new ideas.

Balancing central steering and tailor-made co-production

With all its diverse topics and domains, this book has also shown that depending on the actor, the option or technology and the context, there is a strong need for 'tailor-made' policies to support efforts required in a specific area or domain. Households and the urban environment have specific, local interests, and applying new technologies or modifying consumption patterns have specific, local requirements. In contrast, the petrochemical industry operates in a global market with fierce competition; it may be served far better by agreements across the sector, worldwide, with allowances for emission trading. In one area, the development of new technologies is a main issue; in another, it is just a matter of implementing schemes and making minor changes in the market place. The key issue here is that specific conditions require specific management and policies in order to be successful. Success also requires a decisive government that is capable of developing, implementing and maintaining such sector-specific policies in close co-operation with relevant actors in the market and civil spheres. 'Decisive government' here is not meant as 'strong' government fiercely applying top-down command-and-control mechanisms, but as a forward-looking government, as stable as the North Star, that persistently shows the direction for development.

Such a general requirement contrasts strongly with some current trends where national governments and industrialists rely more and more on market mechanisms and put more and more trust in generic policy instruments (such as tax measures and emission trading). Related to this point, we have concluded that emission trading will certainly play an important role in the (inter)national arena (as is also argued by Woerdman et al. in Chapter 9) but it might be insufficient in itself to ensure that the major transitions required to meet long-term objectives take place. Emission trading (or 'flexmechs') should therefore be part of a balanced mix of policy instruments and strategies, on both the international and the national level.

Handling conflict and equity

The COOL dialogue discussed in Chapter 11 showed that, in the future, controversies may arise around some major technologies necessary for the transition to a climate-neutral society, such as the application of CO_2 storage, biomass and wind, since there will be some groups of actors who oppose them. Controversy and conflict can also grow – or may already exist – between different technological approaches. The visions in Chapter 2 have shown that it is possible to have competing technological routes for reaching the same objective. Here, we have stressed the virtues of diversity and competition, but one should also be aware of the drawbacks arising from controversy and conflict between the visions and among the various actors and sectors.

Chapter 4 by Arentsen et al. illustrated the difficulties of managing technology development and technological change. It is tempting to think that the climate problem can easily be solved by technological means, but the requirements for actually having the essential technologies available at the right time are complex. Arentsen et al. have highlighted the possibility of more coherent technology management over a prolonged period of time. Conflicts will arise between different technologies and approaches, but these frictions may also serve as a motivation for change. In policy and technology management, explicit attention should be paid to this. Risks are involved in developing alternatives, and a long-term perspective and stable long-term policies are essential. It is unlikely that market parties will accept all the risks in developing options, infrastructure and the like. Therefore, a key role is identified for (supra-)national governments in creating the right conditions for technological change and development. Inevitably this will involve high costs in the early years. On the other hand, improved international collaboration and stable long-term policies may yield considerable efficiency benefits in the long term.

Another source of conflict lies in equity. The burdens (or benefits!) of transitions will not be distributed evenly throughout society; neither will the efforts essential to obtaining change and activity be equally distributed throughout sectors and domains. Many actors depend on each other for success, and in order to keep their efforts focused on long-term goals, dialogue and debate are essential on many levels: sometimes local, sometimes national, and often international. There are very positive examples of such dialogues – which can serve as models or starting points for dialogues about long-term climate objectives – such as the COOL dialogue discussed above, the protocol of Montreal on reducing the emissions of CFCs to protect the ozone layer (a so far successful example of an international agreement on a global environmental problem!) and the agreement in the world's adipic acid industry to join efforts to develop and implement new technologies that phase out the significant N_2O emissions in this sector (Reimer et al., 2000). In fact, the global climate conferences (Conferences of Parties) are a spectacular result of international dialogue on a global issue in their own right, uniquely resulting in agreements among a large number of countries. Much is to be improved in this type of process, but dialogue has already led to consensus and an understanding of the problems and the need to deal with them. Such examples show that dialogue about complex problems and long-term objectives is possible and can yield good results.

The authors of this book were asked to go beyond the borders of their scientific expertise in a challenging exercise aimed at shedding light on issues related to 'trend breaking' from a scientific perspective. An understanding of the science and practice of trend breaking or transition management will be essential if substantial reductions in greenhouse gas emissions are to be realized in the long term. Yet, there is little experience in this field and clear itineraries are not yet available. We hope that this book will inspire you to make your own contribution to the art of trend breaking.

References

Driessen, P.P.J. & P. Glasbergen (eds.), *Greening Society. The Paradigm Shift in Dutch Environmental Politics*, Kluwer Academic Publishers, Dordrecht, 2002.

Glasbergen, P. (ed.), *Co-operative Environmental Governance. Public-Private Agreements as a Policy Strategy*, Kluwer Academic Publishers, Dordrecht, 1998.

Ministerie van VROM (Ministry of Environment), *Milieuprogramma 2001-2004* [Environmental Program 2001-2004]. TK 2000-2001, 27404(2), Government of The Netherlands, The Hague, 2000.

OECD, *Sustainable Development, Critical Issues*, Paris, 2001.

Reimer, R.A., C.S. Slaten, M. Seapan, T.A. Koch and V.G. Triner, Adipic Acid Industry – N_2O Abatement. Implementation of Technologies for Abatement of N_2O Emissions Associated with Adipic Acid Manufacture. In: J. van Ham et al. (eds.), *Non-CO$_2$ Greenhouse Gases: Scientific Understanding, Control and Implementation*, Kluwer Academic Publishers, Dordrecht, 2000.

Vermeulen, W.J.V. and R.A.P.M. Weterings, Extended Producer Responsibility: Moving from End-of-Life Management towards Public-Private Commitment on Life Cycle Innovations of Products. *Journal of Clean Technology, Environmental Toxicology and Occupational Medicine* 6, 1997, pp. 283-298.

Vermeulen, W.J.V., Greening Production as Co-Responsibility. In: Driessen, P.P.J. & P. Glasbergen (eds.), *Greening Society. The Paradigm Shift in Dutch Environmental Politics*, Kluwer Academic Publishers, Dordrecht, 2002.

List of contributors

Dr. Henk Addink is associate professor at the Faculty of Law of Utrecht University (the Netherlands). E-mail g.addink@law.uu.nl

Dr. Maarten Arentsen is managing director and senior researcher at the Center for Clean Technology and Environmental Policy of the University of Twente. E-mail: m.j.arentsen@cstm.utwente.nl

Dr. Jan-Tjeerd Boom is assistant professor at the Unit of Economics of the Royal Veterinary and Agricultural University in Denmark. E-mail: jtb@kvl.dk

Dr. Frans H.J.M. Coenen is senior research associate at the Centre for Clean Technology and Environmental Policy (CSTM) at the University of Twente. E-mail: f.h.j.m.coenen@cstm.utwente.nl

Dr. Robbert van Duin is consultant at Bureau B&G, specialized in waste management and recycling. E-mail: Bureau.BenG@wxs.nl

Dr. Andre Faaij is associate professor at the Department of Science, Technology and Society of the Utrecht University. E-mail: A.Faaij@chem.uu.nl.

Rob Folkert is researcher at the National Institute for Public Health and Environment (RIVM), Air Research Laboratory. E-mail: Rob.Folkert@rivm.nl

Dr. Peter Groenewegen is associate professor of organization theory, Faculty of Social Cultural Studies, Department of Public Administration and Communication Science, Vrije Universiteit of Amsterdam. E-mail: p.groenewegen@scw.vu.nl

Dr. Ans Groot-Marcus is assistant professor at Wageningen University and Research Centre, Consumer Technology and Product Use. E-mail: ans.groot@tech.hhs.wau.nl

Dr. Marko Hekkert is assistant professor at the Copernicus Institute for Sustainable Development and Innovation, Department of Innovation Studies at Utrecht University. E-mail: M.Hekkert@geog.uu.nl

Dr. Matthijs Hisschemöller is senior researcher at the Institute for Environmental Studies at the Vrije Universiteit of Amsterdam. E-mail: matthijs.hisschemoller@ivm.vu.nl

Ir. D. de Jager is head of the Energy and Environmental Policies department at Ecofys in the Netherlands. E-mail: d.deJager@ecofys.nl

Dr. Paulien de Jong is researcher at Social Aspects of Science and Technology, Faculty of Sciences, Vrije Universiteit in Amsterdam. E-mail: pdejong@nat.vu.nl

Dr. René Kemp is senior research fellow at Maastricht Economic Research Institute on Innovation and Technology (MERIT). E-mail: r.kemp@merit.unimaas.nl.

Marleen van de Kerkhof is a Ph. D. student at the Institute for Environmental Studies in the field of participatory integrated assessment at the Vrije Universiteit of Amsterdam. E-mail: marleen.van.de.kerkhof@ivm.vu.nl

Marcel Kok, worked as a programme officer at the Dutch National research Programme on Global Air Pollution and Climate Change and now works as a researcher at the National Institute for Public Health and Environment (RIVM), Global Environmental Assessment Division. E-mail: marcel.kok@rivm.nl

Ir. Tom Kram was senior project manager at the Policy Studies Unit of the Netherlands Energy Research Institute. Recently he joined RIVM, Global Environmental Assessment Division. E-mail: tom.kram@rivm.nl

Marc van Lieshout is working as a senior-researcher and advisor at TNO 'Strategy, Technology and Policy'. His main field of interest is ict-innovation processes in the context of the sustainable information society. E-mail: vanlieshout@stb.tno.nl

Dr. Esther Luiten performed her PhD research at Utrecht University, department of Science, Technology and Society. She is currently working at the Netherlands Study Centre for Technology Trends (STT in The Hague. E-mail: luiten@stt.nl

Ir. Marijke Menkveld is a scientist at the Policy Studies unit of the Netherlands Energy Research Centre (ECN), Petten. E-mail menkveld@ecn.nl

Dr. Henk Moll is associate professor at the University of Groningen, Center for Energy and Environmental Studies. E-mail: h.c.moll@fwn.rug.nl

Dr. Andries Nentjes is professor at the Department of Economics and Public Finance (ECOF) within the Faculty of Law of the University of Groningen in the Netherlands. E-mail: a.nentjes@rechten.rug.nl

Adriaan Slob is manager of the departement 'Sustainable Development' within TNO Strategy, Technology and Policy. E-mail: slob@stb.tno.nl.

Dr. Walter Vermeulen works as environmental sociologist at the Copernicus Institute for Sustainable Development and Innovation of the Utrecht University. E-mail: w.vermeulen@geog.uu.nl

Edwin Woerdman is researcher at the Department of Economics and Public Finance (ECOF) within the Faculty of Law of the University of Groningen in the Netherlands. E-mail: e.woerdman@rechten.rug.nl

Index